H. Rawlinson
Associate of the Chartered Institute of Building

Site surveying and levelling
Level 2

Longman London and New York

Longman Group Limited

Longman House
Burnt Mill, Harlow, Essex, UK

Published in the United States of America
by Longman Inc., New York

© Longman Group Limited 1982

First published 1982

British Library Cataloguing in Publication Data
Rawlinson, H.
    Site surveying and levelling, level 2. —
    (Longman technician series: construction
    and civil engineering sector)
    1. Surveying   2. Levelling   3. Building sites
    I. Title
    624   TA5454

    ISBN 0-582-41597-7

Library of Congress Cataloging in Publication Data
Rawlinson, H., 1931—
    Site surveying and levelling

    (Longman technician series. Construction and
civil engineering sector)
    Includes index.
    1. Surveying.   2. Leveling.   I. Title.   II. Se-
ries.
TA545.R18   526.9   81-8122
ISBN 0-582-41597-7   AACR2

Typeset by Lonsdale Typesetting Services and
Printed in Great Britain by
Butler & Tanner Ltd, Frome and London

Longman Technician Series

Construction and Civil Engineering

General Editor — Construction and Civil Engineering

C. R. Bassett, B.Sc.
Formerly Principal Lecturer in the Department of Building and Surveying, Guildford County College of Technology

Books already published in this sector of the series:

Building organisation and procedures   *G. Forster*
Construction site studies — production, administration and personnel
   G. Forster
Practical construction science   *B. J. Smith*
Construction science Volume 1   *B. J. Smith*
Construction science Volume 2   *B. J. Smith*
Construction mathematics Volume 1   *M. K. Jones*
Construction mathematics Volume 2   *M. K. Jones*
Construction surveying   *G. A. Scott*
Materials and structures   *R. Whitlow*
Construction technology Volume 1   *R. Chudley*
Construction technology Volume 2   *R. Chudley*
Construction technology Volume 3   *R. Chudley*
Construction technology Volume 4   *R. Chudley*
Maintenance and adaptation of buildings   *R. Chudley*
Building services and equipment Volume 1   *F. Hall*
Building services and equipment Volume 2   *F. Hall*
Building services and equipment Volume 3   *F. Hall*
Measurement Level 2   *M. Gardner*
Structural analysis   *G. B. Vine*

Contents

Preface

This elementary surveying book requires no previous knowledge of the subject. It has been specifically prepared to cover the topics in the Technical Education Council (TEC) Level II standard unit 'Site Surveying and Levelling'. It will be the student's introduction to the subject as surveying does not occur at the TEC Level I stage. The book does not cover any topics in the more advanced Level III standard unit 'Site Surveying', but does deal with 'areas' which are not shown in either unit.

To give the students a clear understanding of the subject the text has been augmented, wherever possible, with explanatory tables, sketches, scale drawings and photographs. Questions at the end of each chapter should be used to test progress. Students should resist the urge to refer to the answer at the back of the book until the question has been completed. The TEC prerequisite unit is 'Mathematics I'. For anyone not following a TEC course an elementary knowledge of trigonometry is useful for parts of the book. All necessary formulae are given in Appendix 1.

The level of the book is perfectly adequate for surveying work involved on small construction sites and for the measuring of existing buildings. It is also suitable for other industries where elementary surveying is carried out, for example landscaping. The book covers the basic surveying techniques on which more advanced work can be built to prepare students for the professional examinations of the construction, engineering, surveying and similar institutions.

Acknowledgements

The author wishes to thank the following companies for providing photographs and information on surveying and draughting instruments: Blundell Harling Ltd. for Fig. 2.32; Hall and Watts Ltd. for Figs. 2.23, 2.26, 5.2, 5.8, 7.2, 7.5, 7.9, 7.10 and 7.11; Rabone Chesterman Ltd. for Figs. 2.3, 2.4, 2.5, 2.7, 6.1, 7.1, 7.6 and 7.8; and Vickers Ltd. for Fig. 5.6. The following bodies have kindly granted the author permission to make reference to and reproduce extracts from their publications: Building Research Establishment for extracts from Current Paper CP15/77 and Digest 114 (Crown copyright); British Standards Institution for extracts from BS 1192:1969 and BS 4484: 1969 (complete copies can be obtained from 101 Pentonville Road, London N1 9ND); Construction Industry Research and Information Association for extracts from *Manual of Setting-out Procedures* published by CIRIA; and The Ordnance Survey for an extract of 1 : 1250 mapping.

The author is indebted to Mr C. R. Bassett, the General Editor of the series for his advice during the preparation of the book, and finally would like to thank his wife, Jean, who typed the manuscript.

Chapter 1

An introduction

1.1. Surveying terms

Surveying

Surveying is the measuring of related existing detail on the earth's surface and subsequently showing this information either by scale drawing to form a map or a plan, or by calculation usually in the form of co-ordinates. The existing detail which is being measured can be naturally formed features such as hills and rivers or man-made features such as roads and buildings. Each item of detail must be related to adjacent detail to form a comprehensive picture of the area.

The difference between a map and a plan is that a plan shows all detail reduced proportionally and is used mostly for development purposes. A map is drawn to a much smaller scale at which it is impossible to show all detail clearly at the same scale. Depending on the purpose of the map, certain detail is emphasized for easy recognition. On route maps, for example, different road classifications are shown by different thickness of lines.

Setting-out

Setting-out is the positioning of predetermined detail on the ground to show the location of proposed construction work both horizontally and vertically.

Levelling

Whereas surveying is concerned only with the horizontal portrayal of information, levelling involves vertical measurements which, when added to the surveyed plan, give a complete picture of the terrain. Levelling is also used during the construction process to establish the depth of drains and foundations and the height of floors and roofs.

Level plane

When finding the height of points on the earth's surface a common level surface reference must be used. A level surface can be defined as being parallel with still water and as water follows the curve of the earth's surface the level surface will also curve. To give an example of the extent of curvature, making allowance for refraction which slightly bends the sight line downwards, a person of normal height sees the level horizon 8 km away from where he is standing. Anything on the surface beyond this distance is not visible. This is illustrated in Fig. 1.1.

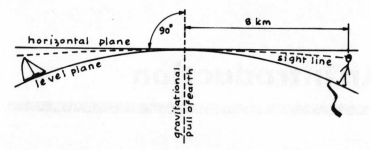

Fig. 1.1 Curvature of the earth

Surveys of very large areas which need to take into account the curvature of the earth's surface are known as *geodetic surveys*. Surveys of smaller areas where curvature can be ignored are called *plane surveys*. In plane surveying the level plane is considered to coincide with the horizontal plane. The horizontal plane is defined as an imaginary flat surface at right angles to the earth's gravitational pull.

Surveying on construction sites is plane surveying.

Height

The height of any object is the vertical measurement from a reference surface to the top of that object. Where the height of an object is given, the reference surface used should be clearly stated. The height of a tree refers to the vertical measurement from the adjacent ground level, to the top. The height of a room is the vertical dimension from the finished floor level to the ceiling, and the height of the ground at any point is the vertical distance of that point above mean sea level. During construction the height of any part of the structure may be related to an arbitrary imaginary level surface below ground level.

Linear measurement

The word linear means 'in-line', a linear measurement being a measurement of length but it does not have to be taken in a straight line. Measurements taken on the centre-line of a road will follow any curves in the road and the girth (circumference) of a tree trunk is also a linear measurement.

Linear measurement can be made in three different ways.

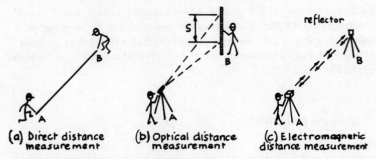

Fig. 1.2 Methods of linear measurement

1. Direct. Made with a tape measure or similar instrument stretched between the points A and B as in Fig. 1.2(a).

2. Optically. An instrument positioned at A reads an instrument at B enabling the length AB to be calculated. Fig. 1.2(b) shows a level at A taking stadia readings on a staff at B. The stadia distance 'S' multiplied by a constant for the instrument (normally 100) will give the length AB.

3. Electromagnetically. Figure 1.2(c) illustrates the distance AB being found by timing electromagnetic radiation transmitted from A to B and back to A.

Angular measurement

Assume that three points A B C are marked with pegs at random on a level field. The surveyor positioned at A directs an instrument's telescope to peg B and then swings it round to observe peg C. The amount that the telescope has moved is part of a horizontal circle and this angular measurement is expressed in degrees and parts of a degree. Instruments used in other parts of Europe may record in grades and decimals of a grade.

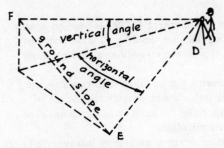

Fig. 1.3 Angular measurements

Vertical angular measurement is used, for example, to find the incline of sloping ground and is the angular difference between the horizontal and the slope line. The instrument will show this measurement in degrees or grades but some instruments also record gradients and percentage slopes.

Consider three pegs D E F positioned on a sloping field so that E and F are at different heights. With the instrument positioned at D the telescope is firstly directed at E and then swung sidewards and also tilted so as to focus on F. In this case both a horizontal angle and a vertical angle will be measured as shown in Fig. 1.3.

1.2.   Surveying principles

Methods

A framework of survey lines is arranged to cover the site and from these lines the detail is measured. Four surveying methods of relating point C to a line AB are described.

1. Trilateration

This is the process of relating the main survey points by linear measurement only. The lengths AB, AC and BC are measured on site as in Fig. 1.4(a) and in chain surveying the direct measurements are drawn to scale as in Fig. 1.4(b) to form a plan.

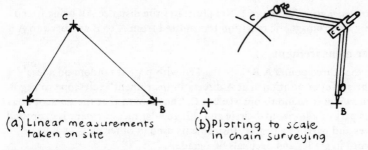

(a) Linear measurements taken on site

(b) Plotting to scale in chain surveying

Fig. 1.4 Trilateration

Where electromagnetic distance measurement (EDM) is used to obtain very accurate measurements over long distance drawing to scale does not provide the necessary degree of accuracy so the co-ordinates of the triangle points are calculated.

Although optical linear measurements can be taken with common site surveying instruments such measurements do not provide the degree of accuracy required for trilateration. Specially designed instruments are used for the purpose and the triangle points calculated.

The operation of EDM and optical distance measuring instruments is beyond the scope of this book.

5

2. Triangulation

This method of surveying relies on a high degree of accuracy in both linear and angular measurements. A base line AB is set up on site. The longer this line is made the greater will be the degree of accuracy possible and the larger the area that can be covered.

The line AB should be located in such a position that, using it as a base, triangles can be measured where required over the site. Once the linear measurement AB is made the angular measurements BAC and ABC are taken as shown in Fig. 1.5(a). Lengths AC and BC are calculated using the sine rule (formulae Appendix 1). From the same base line other base angles for triangles ADB and AEB can be measured (Fig. 1.5(b)) and the sine rule used to calculate lengths AD, BD, AC, BC, AE and BE. Lengths DC and CE are found by the cosine rule.

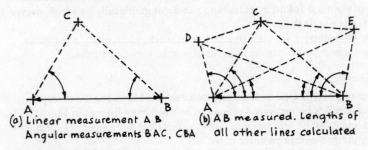

(a) Linear measurement AB
Angular measurements BAC, CBA

(b) AB measured. Lengths of all other lines calculated

Fig. 1.5 Triangulation

The Ordnance Survey used triangulation to map the British Isles using a base line on Salisbury Plain of 11.26 km and one in Highland Scotland of 24.83 km and from these lines built up triangles covering the whole country.

This method is particularly suitable for surveying a site where the ground is rough with intervening obstacles such as pits, a valley or a main road.

3. Polar co-ordinates

This is also known as the radiation method and locates points by horizontal angle and distance. Figure 1.6(a) shows the angular measuring instrument set up on site over a selected point B with the telescope zeroed on another selected point A. The telescope is swung in a clockwise direction to focus on point C and horizontal angle ABC obtained. The horizontal lengths AB and CB are measured on site. To prepare a scale drawing of these points the angle ABC is plotted using a protractor and the lengths AB and BC drawn to scale. This method is useful when surveying detail. The points where the instrument is set up must be carefully related by measured triangles or by polar co-ordinates. Figure 1.6(b) shows detail being surveyed by polar co-ordinates.

The degree of accuracy which can be achieved is in the region of 1 : 500 which is dictated by the plotting of angles with a protractor. It is unlikely that angles can be plotted closer than 5 minutes so there is nothing gained by measuring angles on site to anything less than this amount.

5

6

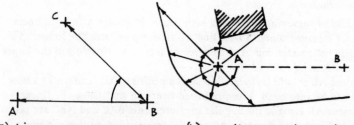

(a) Linear measurements AB, BC
Angular measurement ABC

(b) Detail surveyed at A with instrument set to zero on B

Fig. 1.6 Polar co-ordinates

Where polar co-ordinates are converted to rectangular co-ordinates the position of points is found by calculation and not graphically so a high degree of accuracy is obtainable.

Polar co-ordinates are used in setting-out work, particularly roadworks. The construction points can be positioned within very fine limits.

4. Square offsets

Whichever method is used to form the framework of lines the usual way to survey the detail is by measured offsets from the survey lines as in Fig. 1.7(a). The right-angle can be estimated for short offsets but for longer ones instruments are used to ensure an angle of $90°$.

Square offsets are used for setting-out purposes as shown in Fig. 1.7(b) with a theodolite or sitesquare forming the right-angle. Hand-held squares used in chain surveying do not provide the accuracy required.

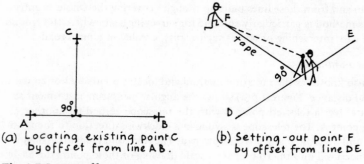

(a) Locating existing point C by offset from line AB.

(b) Setting-out point F by offset from line DE

Fig. 1.7 Square offsets

Control

Survey control keeps the work in check by reference to temporary or permanent points. These control points can either be fixed at random and their position ascertained, or located at a predetermined spot. Horizontal control points are fixed when plan surveys are being undertaken and vertical control points are fixed for height reference.

The Ordnance Survey have established control points throughout Great Britain which can be used by the public. The national grid co-ordinates of each horizontal control point can be obtained from the Ordnance Survey Offices and the heights of vertical control points are shown on the large-scale Ordnance Survey (OS) maps.

Only very large developments would tie-in to the OS grid, the normal construction site would use its own referencing system. The various horizontal control points established during the orginal survey of a site are often left in position to provide initial control points for subsequent setting-out work. Other control points are then fixed on site from which horizontal setting-out measurements are taken. Pegs, rails and painted marks on walls are used on site for vertical control of construction work.

Working from the whole to the part

Consider that a survey is required of a large area. If a method is adopted of starting at one corner and working outwards adding detail as the survey progresses until the whole area has been covered, any small errors would tend to build up, and the error for the whole area could be considerable.

To reduce the effect of such cumulative errors the first stage of surveying is to establish a framework of spaced control points over the site which are measured and checked with the accuracy demanded for the purpose of the survey. If adjacent control points are joined up a series of triangles is formed and detail is added within each triangle. By this method any inaccuracies are confined within each triangle and not allowed to accumulate.

The expression 'working from the whole to the part' could be rewritten as 'working from the initial overall framework to the small detail'.

Accuracy

The definition of accuracy is the closeness of the measured value to the true value. The difference between these two values is the error. The closer the measured value to the true value the smaller the error. No surveying measurement can be claimed to be completely accurate and a surveyor aims at a certain degree of accuracy appropriate to the work being undertaken.

Generally the degree of accuracy of a survey cannot be found because the true value is unknown and the error cannot therefore be established. However, experience has shown that by using specific surveying techniques and instruments a certain degree of accuracy can be obtained.

The degree of accuracy expected for a small housing site would not normally be more than 1/500. This means that a surveyed line 100 m long should be within 0.2 m of its true length, 0.2 being the maximum error. Greater accuracy is demanded for suveys in town centres and for large area development where errors can become appreciable. A degree of accuracy of 1/2000 is normally acceptable for such surveys giving the maximum error of 0.05 m on a line 100 m long.

The higher the degree of accuracy required the greater will have to be the care and time taken and the more expensive will be the equipment necessary to achieve the result. It is essential that the degree of accuracy of site work

8

suits the purpose of the required survey. Where a survey will be presented as a drawn plan, without controlling measurements, there is no point in working to a higher degree of accuracy in the field than can be plotted with drawing instruments in the office.

Precision

When applied to surveying the word precision has two distinct meanings but in neither case is it a direct indication of accuracy.

1. *In mathemetical terms* precision indicates the spread of measured values. Values grouped closely together are said to be more precise than those more widely spaced.

A surveying example is of two different measuring tapes used several times to mark off the same length from a common origin. When using the first tape the marks happen to occur closely together but the tape measure was found to have stretched. A second tape, checked and found to be correct, is used but the measured marks are more widely spaced than before. Comparing the two results the first tape gave a more precise result than the second but the second gave a more accurate result than the first. This is illustrated in Fig. 1.8.

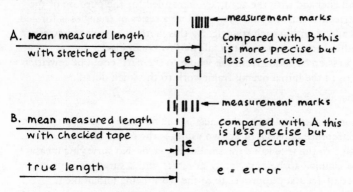

Fig. 1.8 Example of mathematical definition of precision

2. *In surveying terms* precision is often used to describe fineness of measurement. Precision instruments are manufactured to closer tolerances than normal models and have added refinements enabling readings to be made to very small limits. Most theodolites read horizontal angles to 20-second divisions but manufacturers refer to precision theodolites as those that read to 1 second or less. The use of such instruments enables the surveyor to take extremely fine measurements but such measurements by themselves do not necessarily constitute work of high accuracy. The surveyor, for example, could be directing the instrument at the wrong points on the site.

It will be seen that precision in both its mathemetical and surveying definitions does not automatically contribute to greater accuracy and will only do so if all other aspects involved are dealt with correctly.

Types of error

Error is the difference between the measured value and the true value. There are three distinct types of error that occur in surveying.

Gross errors

These are human blunders caused by inexperience or carelessness. They are caused by such actions as miscounting, misreading and wrong booking. Such mistakes are bound to occur occasionally so all stages of the fieldwork, plotting and calculations must be provided with means of checking so that gross errors can be detected and eliminated. Main site measurements should be checked by independent measurements, by remeasurement, by calculation or by graphical means.

Systematic errors

These are constant errors that can be calculated and so corrected. They are referred to as cumulative errors as they increase at a constant rate. Systematic errors in linear measurements can be caused by incorrect instruments, and the effect on the measuring instruments due to, say, tension and temperature. Angular and levelling systematic errors are most likely to be caused by badly adjusted instruments. Such errors can be reduced or completely eliminated by applying the principle of reversed readings which cancel out several instrumental errors.

All instruments should be checked for accuracy and corrected at frequent intervals.

Compensating errors

These are unavoidable random errors caused by the fact that a human observer cannot repeat certain procedures exactly. If a length is measured several times under identical conditions the readings are likely to vary each time by a very small amount. The same applies to angular and level readings and also to plotting.

In practice it is found that larger readings will occur at the same frequency as smaller ones so, if sufficient number of readings are taken, discrepancies tend to cancel out and for normal surveying work compensating errors can be ignored completely.

Checking

Mistakes of various magnitude are bound to occur from time to time in any discipline undertaken by man and surveying is no exception. When surveying mistakes go undetected until the plan is being used for development purposes the result can cause delay, confusion and usually financial loss. The simplest remedy may be a re-survey but if construction has already started serious problems can develop, in extreme cases necessitating demolition of new work. The importance of checking or proving surveying work at all stages cannot be over-emphasized.

It is not feasible or necessary to check small detail work but the main framework must be verified together with important detail. An independent checking method should be introduced wherèver possible. Failing this, repeat measurements of the required work should be made. The work should be divided into sections for checking purposes so that faults can be isolated. Verifying that the survey measurements are correct should be done on site wherever possible so that discrepancies can be found and dealt with straight-away. Other errors will not show up until the plotting or calculation stage in the office.

1.3. Scales

To plot a survey means to draw it proportionally on paper or other draught-ing material to scale. The ratio of the detail on site to the same detail on the drawing is the scale expressed as a representative fraction. A scale of 1 : 1000 signifies that site detail is drawn at 1/1000th of its true size on paper. Metric scale rules are available in lengths of 300 mm office size and 150 mm pocket size. They each have two or more scales recommended in BS 1347:Part 3: 1969. The most useful single all-purpose scale rule for surveying has scales of 1 : 1, 1 : 100, 1 : 20, 1 : 200, 1 : 5, 1 : 50, 1 : 1250, 1 : 2500, the last two for use with OS sheets. Other scales will be used when plotting, for example 1 : 500. The 1 : 50 scale can be adapted for 1 : 500 work by mentally adding a zero on to each figure. A surveying office would have a comprehensive supply of scale rules covering all the recommended scales.

The scale chosen for plotting a particular survey is dependent on the purpose of the survey and to some extent on the size of paper to be used. It is always better to show a complete survey on a single sheet where possible. Table 1.1 shows appropriate scales for drawn plans.

Table 1.1 Scales for drawn plans

Type of drawing	Preferred scales BS 1192:1969	Statutory requirements Building Regulations
Location or key plan	1 : 2500	Not less than 1 : 2500
Block plan	1 : 1250	Not less than 1 : 1250
Site plan	1 : 500, 1 : 200	
General location	1 : 200, 1 : 100	
	1 : 50	

On occasions it may be necessary to work from old plans drawn in imperial scales so a metric scale rule is of no use. Metric scales can be pre-pared to suit the imperial drawings as shown in Fig. 1.9.

Example : To prepare a metric scale to suit a plan drawn at 1 inch to represent 16 feet.

Imperial	1 inch represents 16 feet
express in mm	25.4mm represents 16 × 304.8 mm
therefore	1 mm represents $\dfrac{16 \times 304.8}{25.4} = 192$

Representative fraction 1:192

Construct the scale of 60 m length

Total length of scale in mm $\dfrac{60 \times 1000}{192} = 312.5$ mm

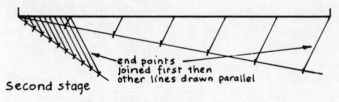

312.5 mm

Suitable length stepped out 10 times

Suitable length stepped out 6 times

First stage

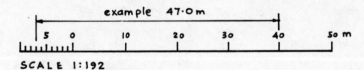

end points joined first then other lines drawn parallel

Second stage

example 47.0 m

5   0   10   20   30   40   50 m

SCALE 1:192

Final stage

Fig. 1.9 Constructing a scale

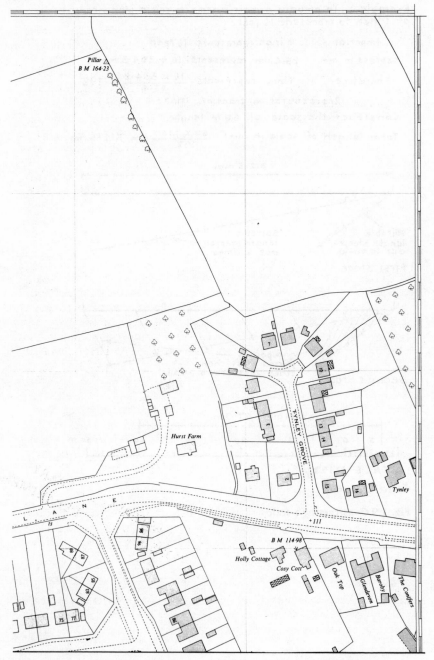

Fig. 1.10 Part of OS Sheet. Scale 1 : 1250 (Crown Copyright Reserved)

Ordnance Survey maps

The Ordnance Survey is a goverment department founded in 1791 whose present charter makes it responsible for the official surveying and mapping of Great Britain. The department produces maps ranging from 1 : 1250 to 1 : 125 000. Large-scale maps of 1 : 1250 and 1 : 2500 are the most useful to the site surveyor. Part of a 1 : 1250 sheet is given in Fig. 1.10. It shows 100 m grid squares, a triangulation station pillar, a bench-mark on the pillar, a bench-mark on a house and a surface level on the centre of a road. All levels relate to Newlyn datum.

Crown copyright subsists in all Ordnance Survey publications and they may not be reproduced either in whole or in part without the permission of the Ordnance Survey. Enquiries should be addressed to Copyright Branch, Ordnance Survey, Romsey Road, Maybush, Southampton, Tel: Southampton 775555.

1.4. Safety

Surveyors have to work on busy roads and construction sites where moving vehicles and plant, together with other site operations, constitute continual safety hazards. Also, the surveyor by his own actions must not jeopardize the safety of others. The surveyor should be familiar with the various government regulations concerning safety both on and off the site. Failure to comply with regulation requirements is an offence. Some of the main safety points are described below.

1. Always be seen
Use fluorescent jackets or reflective belt on public roads or site haul roads. Set up surveying warning triangles on public roads well in advance of the working area. The minimum size of triangle and distance in advance depends on the average speed of private cars using the road. On site make sure that plant operators know where surveyors are working.

2. Protective clothing
Use safety helmets where overhead work is being done, remembering that bright colours are more easily seen. Safety boots incorporating a steel plate in the sole should be worn as protection against projecting nails where timber formwork is being struck.

3. Electrical hazards
Before working in the proximity of exposed electrical apparatus it should be isolated. Particular danger occurs with metal or wet wooden levelling staves used near overhead power lines. High voltage current arcs across wide gaps.

4. Particular situations
(a) Linear measurements across busy roads or electrified rails should be made using optical or electromagnetic means.

(b) Work in deep trenches or tunnels should only be done where adequate precautions have been taken against collapse.
(c) To prevent back injury when lifting manhole covers the back must be kept straight and lifting done with the legs.
(d) Where manhole covers have been removed a guard frame must be placed around the opening.

1.5. Questions

Formulae shown in Appendix 1 (p. 154).
Qu 1.1 Applying the method of triangulation calculate the lengths AC, BC, AD, BD, CD shown in Fig. 1.11 (use sine rule and cosine rule).

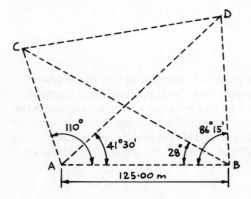

Fig. 1.11

Qu 1.2 Polar co-ordinates are used to set-out the external corners of a nonagonal building from corner A on Fig. 1.12. Calculate the polar co-ordinates from line AB to points C, D, E, F, G, H, and J (use sine rule).

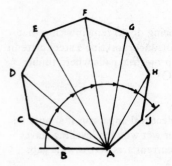

Length of sides = 10·0 m

Note :
In a regular polygon where lines radiate from one corner to all other corners the angle between the radial lines is constant.

Fig. 1.12

Qu 1.3 Houses are to be set-out by square offsets from base line A—B in Fig. 1.13. Calculate the offset lengths to corners D, E, F, and the running dimensions (all measured from A) to the offset points H, J, K, L.

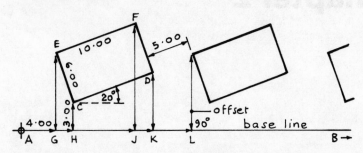

Fig. 1.13

Qu 1.4 (a) If an error of 125 mm is found on a line measured at 150 m what degree of accuracy has been achieved?

(b) If the degree of accuracy of 1/2000 is specified for a survey, what will be the permissible error on a line 275 m long?

Qu 1.5 A metric scale is to be produced to use on an imperial scale map of 6 inches to represent 1 mile. Calculate the representative fraction and use it to construct a metric scale of 3 km total length (6 inches = 152.4 mm, 1 mile = 1609.334 m).

Answers shown in Appendix 2 (p. 156).

Chapter 2

Chain surveying

This is the simplest form of surveying and uses inexpensive equipment. The surveying method used is trilateration. The linear measurements of the triangular framework can be made with land chain or tape, in either case they are called chain lines. Site details are 'picked up' by means of offsets measured from the chain lines. The triangles are plotted on paper or film with compasses and then the detail is added.

Chain surveying is particularly suitable for small open sites where triangles can easily be formed and for that reason is not suitable for town work. While it is difficult to lay down hard and fast rules it is suggested that sites of more than 2 hectares require a more accurate method of framework control. The use of offsets for 'picking up' detail is also employed in other surveying methods. Normally a degree of accuracy of 1/500 should be achieved but with careful work 1/1000 is possible.

2.1. Instruments for measuring and marking

Land chain

Made from galvanized iron wire or black enamelled steel wire which has been hardened and tempered.

Metric chains conforming to BS 4484:1969 have 100 links and are 20 m long including the swivel handles. The arrangement of links is shown in Fig. 2.1. Unnumbered yellow tally markers are connected at each whole metre position. Red tally markers 5 m from each end are marked with the figure 5 and at the centre of the chain a red tally shows the figure 10.

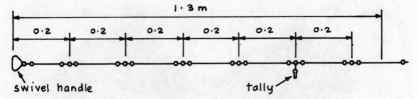

Fig. 2.1 Chain links

A chain is robust and can be dragged over rough ground without harm. Its weight allows it to stay in place on the ground while offsets are being measured.

Periodially a chain should be examined to close any open links and straighten any bent ones. At the same time its length should be standardized (standardizing is explained later).

Any inaccuracy in the length is likely to be stretch and can be remedied by shortening or removal of some of the small connecting links at regular spacings.

To use the chain remove the strap and, holding the handles in the left hand, cast out the rest of chain with the right hand. The chain can now be stretched in a straight line by gently shaking out any tangled links. When gathering in the chain place the handles together on the ground and walk to the 10 m tally marker. Place each set of four links in the palm of the left hand, the links to be slightly inclined to the ones already held as Fig. 2.2. By keeping the links twisting in the same direction the gathered chain takes the shape of a wheatsheaf and the strap can be fastened. Figure 2.3 shows a photograph of land chain.

A muddy chain should be washed in water and dried with a cloth.

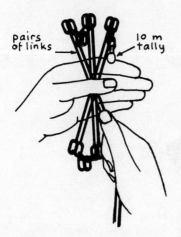

Fig. 2.2 Gathering in the chain

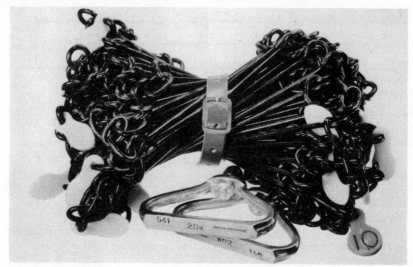

Fig. 2.3 Land chain

Fig. 2.4 Steel band

Band chain and steel band

The British Standard band chain is a steel tape not less than 16 mm wide and 20 m long wound on an open frame. It is removed from the frame for use and its total length is inclusive of handles. Each metre division is marked with a brass plate and each 200 mm with a brass stud. The first and last metres have brass studs marking each 10 mm.

A steel band shown in Fig. 2.4 is a tape up to 100 m long on an open winding frame and from which it is detachable. Handles can be fitted to each end but they are not included in the length. BS 4484:1969 specifies that the enamelled or etched surface shall be graduated at 5 mm intervals except the first and last metres which are to show 1 mm graduations.

The band is lighter than the land chain and a higher degree of accuracy is obtainable. It should be standardized at intervals depending on frequency of use. Great care must be taken not to damage the tape. Never tread on it nor allow vehicles to run over it.

After use the enamelled band should be cleaned with a damp cloth but the etched band should be wiped with an oily rag to prevent the bright markings rusting.

Coated glass-fibre tape

The measuring tape is made from strands of glass-fibre coated with PVC. It is wound into a metal or leather case (Fig. 2.5) and if the tape should be

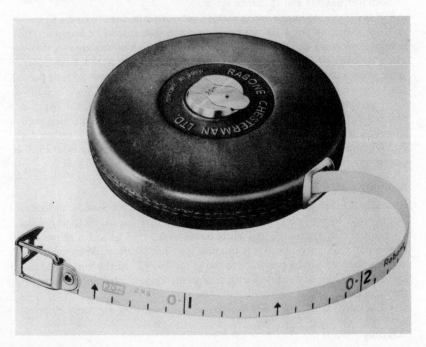

Fig. 2.5 Glass fibre tape

damaged a replacement can be fitted to the case. The tape lengths of 10 m, 20 m and 30 m include the metal finger loop at the beginning. The operating tension is shown on the tape. BS 4484:1969 requires that the tape shall be graduated at 10 mm intervals throughout its length. The length of the tape should be verified from time to time.

To avoid breakage the tape should not be suddenly jerked. Constant winding and unwinding when the tape is in use can be avoided by carefully looping it in the left hand, avoiding tangles. The tape should be cleaned with a damp soapy rag.

Standardizing

This is the process of checking the length of linear measuring instruments against a standard length. The standard length can be a steel tape kept exclusively for checking purposes or a marked length on a footpath or on the floor of a building. Most instruments will stretch with wear and if incorrectly repaired may be shorter than the standard length.

Where a steel tape is used for checking purposes the tension printed on the tape must be applied with a spring balance tension handle with roller grip to attach to the tape. Where spring balances are graduated in kgs 1 kgf = 9.81 newtons. When the checking tape is used to verify the length of a chain, band or other steel tape they should be laid out together for a while to reach ambient temperature. The coefficient of linear expansion for steel will apply to both instruments and so can be ignored.

If the checking tape is being used to set-out permanent markers to be used for checking purposes temperature correction should be applied. The temperature at which the tape was graduated is shown on the tape. A thermometer is positioned next to the tape and the temperature noted. Expansion and contraction cause movement of 0.01 mm per metre length of tape per °C. Most tapes are calibrated at 20 °C.

Example: If a 30 m tape calibrated at 20 °C is used at 2 °C the contraction is 30 x 18 x 0.01 mm = 5.4 mm, which must be added to the measured length of 30 m to give the true length.

Where an invar steel tape is used the temperature movement is so small it can be ignored.

Ranging poles

Normally 2 m long made in glass fibre, or wood with a metal shoe, or in metal throughout. Jointed poles are made to fit into a carrying bag. Ranging poles are painted in alternate red and white bands of 0.5 m. Figure 2.6 shows a tripod stand designed to support the pole on a hard surface. When not is use ranging poles should be pushed into the ground at 45° to the horizontal so that they will not be confused with station markers. If left lying in long grass they can easily be lost.

The only maintenance necessary is repainting if they are badly chipped.

Arrows

Made from heavy gauge hardened and tempered steel wire 400 mm long in

Fig. 2.6 Ranging pole in tripod stand

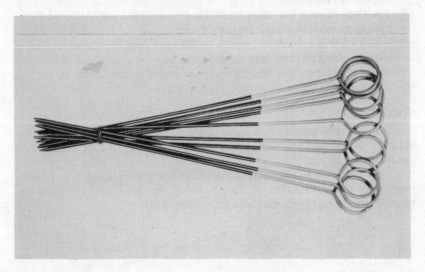

Fig. 2.7 Arrows

sets of 10 (see Fig. 2.7). They should either be painted a fluorescent colour or brightly coloured pieces of cloth should be tied to the ringed tops for easy sighting.

2.2. Chain lines

Main chain lines

These are positioned on the site to form connecting triangles. To ensure accuracy in plotting the arcs struck from the base line should cross distinctly. Such triangles are called 'well-conditioned'. Triangles having angles of less than $30°$ are 'ill-conditioned'. This is shown in Fig. 2.8. The corners of each triangle are called stations and are marked in the field by ranging poles.

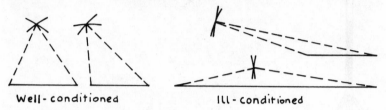

Well-conditioned   Ill-conditioned

Fig. 2.8 Shape of drawn triangles

Principles to be observed when selecting chain lines:

(a) A long line called a base line is established on which the other triangles are based. This acts as a control on the framework of lines.

(b) The lines should avoid obstacles wherever possible.

(c) The lines should be close to the detail being 'picked up' to keep offset lengths to a minimum.

(d) The stations on a line must be visible one from the other so that a straight line can be ranged between them.

(e) Where a line is extended beyond a station, and does not form part of a triangle, this extension line should be as short as possible. If the line needs to be long a new triangle must be formed to incorporate it.

(f) All lines should be within the boundary of the site.

Check lines

To verify the accuracy of chain line measurements taken on site, and also the accuracy of plotting in the office, chain lines acting as checks are introduced and measured on site.

Each triangle is to have at least one check line passing through it. Where possible use existing stations. If triangles are checked separately any error can be attributed to one particular triangle, see Fig. 2.9.

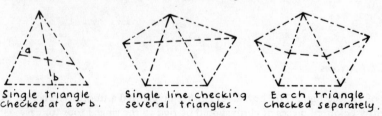

Single triangle checked at a or b.   Single line checking several triangles.   Each triangle checked separately.

Fig. 2.9 Check lines

Detail lines

These are chain lines which are additional to the main framework of lines and are introduced specifically to 'pick up' site detail. Use existing stations wherever possible. Detail lines can also serve as check lines. Figure 2.10 illustrates the application of different types of chain line.

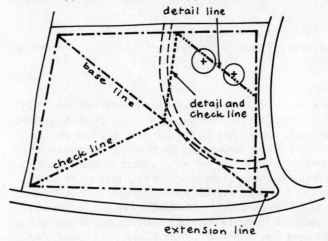

Fig. 2.10 Types of chain line

Witnessing stations

If stations are to be used again, either because the survey cannot be completed in one operation, or because the stations form control points for subsequent work, each must be located by measurement from existing detail. Figure 2.11 shows that care should be taken to set out these stations for easy reference.

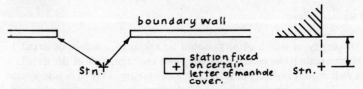

Fig. 2.11 Witnessing stations

Degree of accuracy of site measurements

The scale used for plotting a survey depends on the purpose for which the plan is being prepared. The scale also determines the degree of accuracy of measurements taken on site. Assume that a point can be plotted to an accuracy of 0.2 mm: at a scale of 1 : 1000 0.2 mm plotted on paper represents 1000 x 0.2 mm or 200 mm on site. There is therefore nothing to be gained by booking measurements smaller than 200 mm on site as they cannot be drawn to scale.

If a scale of 1 : 500 is used 0.2 mm on paper represents 500 x 0.2 mm or 100 mm on site which is the required degree of accuracy for site measurements. If a scale of 1 : 200 is used 0.2 mm x 200 = 40 mm. In this case 50 mm is the nearest division to be used.

2.3. Offsets

Measurements are taken off the chain line in one of the three ways described below.

1. Square offsets

A square offset is a measurement at right-angles to the chain line (Fig. 2.12). It is measured with the tape held horizontally between the site detail and the chain. As the right-angle is estimated by eye any error will increase with length. These errors are of little consequence where offsets are taken to ill-defined detail such as a hedge or track but where offsets are taken to a wall, for example, the displacement on a long offset is unacceptable. The actual displacement when plotted varies with the scale being used. As a guide, where surveys are to be plotted to a scale of 1 : 1000, offsets to clearly-defined detail should be limited to 10 m, at a scale of 1 : 500 the limit should be 5 m and at 1 : 200 the limit should be 2 m. Longer offsets can be used when instruments are used to set-out the right-angle or where an arc is swung from the detail to the chain line and the distance between the two crossing points bisected as shown in Fig. 2.12.

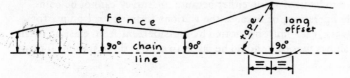

Fig. 2.12 Square offsets

The frequency at which offsets should be taken to continuous detail depends on the scale to be used for plotting and the regularity of the detail. A straight wall requires offsets only at each end. A curved fence beside a main road will be of large radius so offsets at 20 m intervals are probably adequate. A curved fence following the road around a sharp corner on a housing estate will have a much smaller radius and offsets will probably be needed at 5 m intervals or less.

2. Oblique offsets or tie lines

Where an important point such as the corner of a building is being 'picked-up' from the chain line, particularly where its distance from the chain line precludes the use of square offsets, its position can be fixed by forming a triangle with its base as part of the chain line and its other two sides oblique to the chain. This is shown in Fig. 2.13. Such a triangle must be well-conditioned.

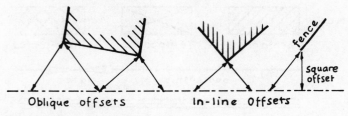

Fig. 2.13 Oblique and in-line offsets

3. In-line offsets

These are measurements from the chain line taken as a continuation of
straight detail. By locating the end of the detail with another offset both the
position and direction of the detail can be fixed as shown in Fig. 2.13.

2.4. Booking field notes

Conventional signs

These should be used when booking in the field and for plotting. See Fig.
2.14.

Field book

A chain survey field book shown in Fig. 2.15 is ruled with red lines to form a
central column. Entries must start at the back of the book and bottom of
the page to give continuity of booking when turning over pages. Notes are
entered in pencil and an eraser carried for corrections. All field notes must be
shown clearly so that there is no ambiguity when plotting is done by the
surveyor or draughtsman.

Notes on booking (illustrated in Fig. 2.16)

(a) State the location of the survey, client and date.
(b) Sketch a key plan of the chain lines showing the stations by letters of the
    alphabet excluding I and O which could be mistaken for figures. Use
    arrowheads to show the direction in which lines will be chained. With a
    clear key plan other lines joining stations need not be shown in the field
    notes.
(c) The central column is used for running chain measurements taken from
    the beginning of the chain line. The figure booked is called the chainage
    at that point.
(d) As the surveyor stands at the beginning of the chain line facing the direc-
    tion of the next station any site detail on his right is booked to the right
    of the central column and detail to his left is booked to the column's
    left.
(e) Detail is sketched in proportion as it occurs on site. Where detail crosses
    the chain it is shown on the appropriate sides of the central column and
    not passing through the column.

26

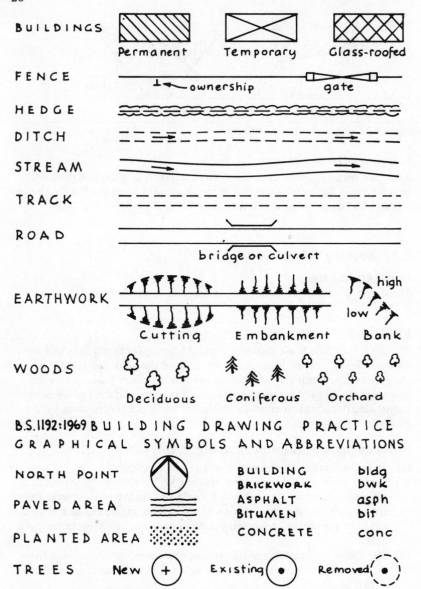

Fig. 2.14 Conventional signs used in surveying

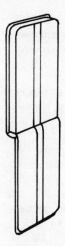

Fig. 2.15 Field book open for note taking

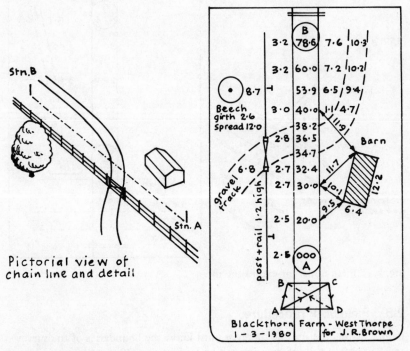

Pictorial view of
chain line and detail

Fig. 2.16 Page of field notes

(f) Square offsets are running measurements between the site detail and the chain, not from one site detail to the next. Dimension lines are not used.

(g) Oblique offsets (or tie lines) and in-line offsets are shown with arrowed dimension lines.

(h) Stations are lettered and circled.

(i) Descriptive notes should show tree species, girth and diameter of branch spread. Fences should show height, construction and ownership, if this can be established.

(j) Each new chain line starts at zero.

(k) Where north cannot be established from an OS sheet the magnetic bearing of a chain line should be found using a prismatic compass.

Booking an extension line

Chain lines can be extended in a straight line beyond the station and can be at the beginning (Fig. 2.17) or end of a chain line. Their purpose is to 'pick up' additional site detail. They are not marked with ranging poles.

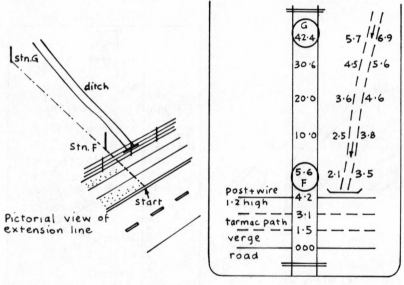

Fig. 2.17 Field notes for extension line

2.5. Fieldwork procedure

Before visiting the site the surveyor must know the boundaries of the survey and quite often this information can be found on an OS sheet.

On arrival the surveyor walks over the site, positioning station ranging poles. The chain lines should follow clear routes such as roads and paths, avoiding where possible dense undergrowth and irregular slopes. When satisfied that the chain lines are correct he sketches the key plan in the field book.

Ideally the surveyor should have two assistants (called chainmen), one to hold the end of the tape while the other calls out the offset chainage, the offset length and if to right or left of the chain line. Where one assistant is used the surveyor has to hold the tape and book the result.

To measure a line the chain is cast out and held by the surveyor against the station ranging pole. The assistant stretches the chain towards the next station and is directed by the surveyor's hand signals. The chain end must be positioned on a straight line between both stations so the assistant should stand to one side while the surveyor sights through. The chain is straightened by 'snaking' as in Fig. 2.18. In short grass the chain end can be seen but in long grass a ranging pole should be held by the assistant for the surveyor's sighting. It may be necessary to cut swathes through undergrowth. When the chain is aligned the end is marked with an arrow from the set.

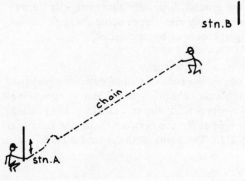

Fig. 2.18 Snaking the chain

Offsets are measured by an assistant holding the end of the tape against the detail while the second assistant, or the surveyor if it is a two-man team, approximates the right angle and reads chainage and offset lengths as illustrated in Fig. 2.19. Offsets to a hedge are measured to its centre-line. A ranging pole should be thrust into the hedge and the offset length found by adding the tape length at the near end of the pole to the length of the pole itself.

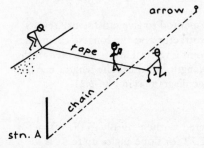

Fig. 2.19 Taking offsets

When all offsets have been taken on the first chain length the surveying team arrives at the 20 m arrow. At this point the leading assistant drags on the chain to measure the next length. At each chain length he places an arrow touching, but not through, the handle and the following assistant collects each used arrow. The number of arrows collected by the follower provides a check on the number of lengths chained.

2.6. Corrections

Slope correction

All maps and plans are plotted in the horizontal plane but field measurements on sloping ground will be longer than the horizontal length so a correction must be applied for plotting purposes. The method adopted depends to a great extent on the nature of the ground slope. For short irregular slopes 'stepping' is most suitable but where long even slopes occur the angle can be measured and correction made graphically or by calculation.

Stepping

This is the simplest method of correction and the chainage shown in the field book is correct for plotting. It relies on a short length of chain or tape being held at an estimated horizontal position while the vertical alignment is made with plumb-bob or ranging pole held at the top between finger and thumb so that it hangs vertically. See Fig. 2.20. The point on the ground is marked with an arrow.

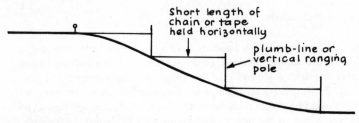

Fig. 2.20 Stepping

Graphical correction

Accurate draughting is required for this method to give satisfactory results.

The slope angle is drawn with a protractor and the longest chain line measured along the slope line. At this point a vertical line is projected to the horizontal line which gives the plotting length. All other site lengths can be converted to plotting lengths on the same diagram as in Fig. 2.21.

Calculated correction

The sloping field book entries are corrected mathemetically in the office for plotting. The correction shown in Fig. 2.22 for slope angles up to 3° is so small that it can be ignored.

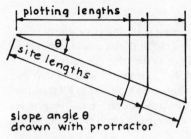

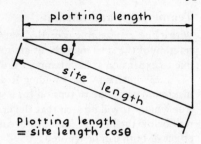

Fig. 2.21 Graphical slope correction

Fig. 2.22 Calculated slope correction

Plotting length
= site length cosθ

Slope angle

In chain surveying the slope angle is found with a clinometer. The most commonly used clinometer is the Abney Level shown in Fig. 2.23. The instrument contains a mirror so that when sighting through the tube the spirit bubble can be seen. The surveyor standing at A sights down the slope to a target at B which is fixed above the ground equal to his eye height. The bubble is centred and the angle noted. The process is now repeated sighting up the slope from B to A and the mean of the two angles used. Vernier scales are used on Abney Levels. The construction and reading of vernier scales is dealt with in Chapter 5 (Figs. 5.3 and 5.4).

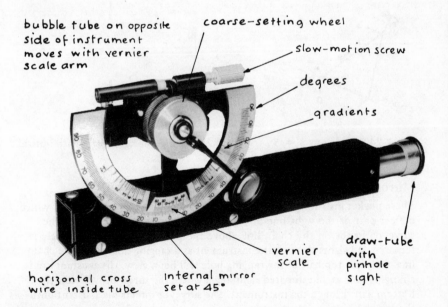

Fig. 2.23 Abney Level

32

Temperature and tension

Where linear measuring is carried out at freezing point the error due to contraction of the steel instrument amounts to 1/5000 of the true length. Error
due to expansion at high temperature will not exceed this amount.

Experiments by the Building Research Establishment indicated that the
maximum error due to tension for a steel supported tape was 1/3000 of the
true length. It will be seen that the errors due to temperature and tension are
well below the normal degree of accuracy expected in chain surveying so the
effect of both can be ignored.

2.7. Setting-out right-angles to chain line

3, 4, 5 triangle

Figure 2.24 illustrates how a single glass-fibre tape can be used for this operation. Place an arrow through the loop end of the tape on the chain line where
the right-angle is to be set-out. Place another arrow 3 m away on the chain
line. The other triangle sides of 5 m and 4 m are added to the 3 m to make
12 m total. The 12 m point is held securely at the first arrow and the apex
point of 8 m (3 + 5) is pulled out tight and a third arrow placed at that point
so forming the right-angle.

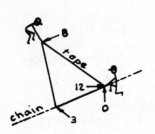

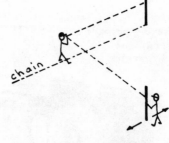

Fig. 2.24 Setting-out 3, 4, 5
triangle

Fig. 2.25 Setting-out right-angle with optical
square

Mirror or prism optical squares

These pocket instruments shown in Fig. 2.26 contain mirrors or prisms which
deflect part of the sight line 90° while still viewing straight ahead. The
surveyor stands on the chain line above the point where the right-angle is
required and sights through the instrument at a ranging pole along the chain
line. Only the top half of the ranging pole will be in view. His assistant holds
a ranging pole at an estimated right-angle where the surveyor will view the
bottom half through the instrument. The surveyor directs his assistant until
the top and bottom images appear as one continuous pole, so forming a right-
angle as illustrated in Fig. 2.25. Alternatively, to locate the point on a chain
line which is set at 90° to a fixed ranging pole the surveyor walks along the

chain line sighting through the instrument until the top and bottom images appear as one.

(a) Mirror type

(b) Prism type

Fig. 2.26 Optical squares

Cross staff

This instrument can be cylindrical or cross arms shape with sights set at $90°$. Its disadvantage is that it is too large to be easily carried in the pocket.

2.8. Obstacles

Where a chain line cannot avoid an obstacle in its path the correct length and alignment can be found by application of simple plane geometry.

Line passing over an obstacle

Figure 2.27 shows a pond obstructing a chain line at A and B. Point C is positioned to one side of the obstacle. Point D is located halfway on line AC and point E halfway on line BC. The length DE is measured which is half of the required length AB.

A building obstructing a chain line

A subsidiary chain line is arranged to run parallel with the main line shown in Fig. 2.28. Great care must be taken with measurements and right-angles. The longer the length of the subsidiary line the greater will be accuracy.

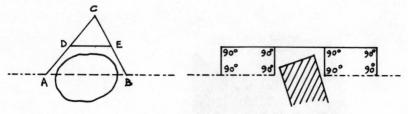

Fig. 2.27 Chain line passing over a pond

Fig. 2.28 Chain line obstructed by a building

Measurement to an inaccessible point

Method A

Locate a clearly definable point A on the opposite side of a river or busy road. On the chain line locate point B which is at $90°$ to A. Position points C and D on the chain line so that BC = CD. The surveyor stands at D and directs his assistant on a line at $90°$ to the chain line. His assistant establishes point E to line up with C and A. Two identical triangles have been formed and the length AB is the same as DE which can be measured (Fig. 2.29).

Method B

Position point L at $90°$ to the chain line at K and at a fixed distance from it. The line MN is also at $90°$ to the chain line and twice the length of KL. If N is made to align with L and J then JK equals KM which can be measured (Fig. 2.30).

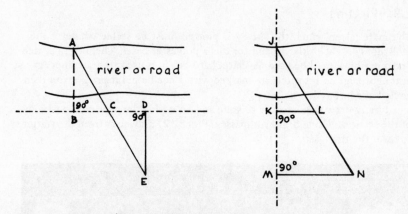

Figs. 2.29 and 2.30 Methods of measuring to an inaccessible point

Ranging a line over a hill or through a depression

The stations A and B in Fig. 2.31 are fixed and the surveyor is required to lay down a line between the two which is straight on plan. Ranging poles C and D are located fairly close together. Standing at D the surveyor directs his assistant to move C to form a straight line DCA. Next his assistant standing at C directs that D is moved to form a straight line CDB. By alternately aligning DCA and CDB a straight line on plan is achieved.

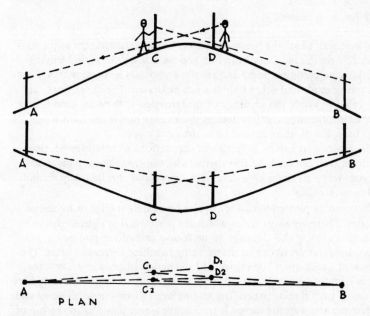

Fig. 2.31 Ranging a line over a hill or through a depression

2.9. Plotting

The material on which the survey is plotted must be stable whatever the moisture content of the air and for this a polyester based, chemically coated drafting film should be used. A sharp, hard pencil is used to draw lines against a straight edge and stations are marked with a needle point pricker and then circled and lettered.

Firstly draw the base line to scale and on it plot the measured positions of the stations. Using beam compasses (Fig. 2.32) draw arcs from the stations to form the triangles.

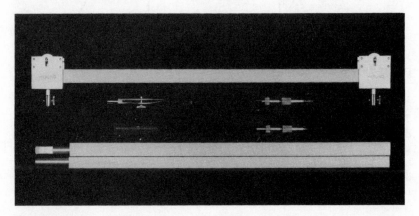

Fig. 2.32 Beam compasses

The best way to set the beam compasses to the measurement required is to draw a line on the bottom of the film and on it mark the scaled lengths measured from a common origin and set the compasses to these marks. From the first triangles plotted other triangles can be drawn. The check lines can now be scaled to verify the accuracy of the triangles. If they are found to be inaccurate the plotting must be checked thoroughly and if the fault is not found to be in the plotting it must be in the field work.

If the faulty part can be isolated and the stations were witnessed, that part of the survey can be done again, otherwise the complete site must be re-surveyed. When all chain lines have been drawn and checked they should be lined-in with red ink.

Offsets can be plotted with a set square on a straight edge or by use of offset scales. These are short scales which are positioned at right-angles to the normal scale enabling the chainage to be found and offset plotted in one operation. Irregular detail can be drawn using flexible or French curves. The destination of roads should be shown and the names of property. Once the drawing is complete it can be traced for printing. Polyester based tracing film is more stable than tracing paper. The tracing must show the north point and this determines the way the survey is traced, the north pointing to the top of the sheet. Chain lines and stations are not shown on the tracing.

Distortion-free prints require a method of reprography where the sheets are laid flat. Where the printing process involves both sheets moving around curved glass some increase in the size of the print is inevitable. The larger the radius of the curve the smaller the distortion. Again the polyester based print gives the most stable results.

2.10. Errors

Gross errors
(a) Misaligning the chain.
(b) Using a chain with tangled links.
(c) Misreading the chain, particularly the 5 m, 10 m and 15 m positions.
(d) Miscounting the number of chain lengths.
(e) Incorrect booking.
(f) Slope allowance not made or miscalculated.
(g) Plotting errors due to misreading poorly written field notes.
(h) Plotting errors due to incorrect scaling.

Remedy
Check at each stage of the work.

Systematic errors
Caused by inaccurate instruments.

Remedy
(a) Chains and tapes. Standardize at frequent intervals.
(b) Squares. Set-out right-angle from both sides of a straight line.
(c) Abney Level. Take readings between two points up and down hill. Error removed by bubble endscrews.

2.11. Example of a chain survey

The area to be surveyed in Fig. 2.33 comprises Alan House and garden with adjacent field. The roads are not to be included and slopes do not exceed $3°$. North was established from a large-scale OS sheet of the area.

There are usually several ways of arranging chain lines on a site. Four different methods have been illustrated in Fig. 2.34. All triangles have check lines and these generally also act as detail lines. Each method is satisfactory but method A has the advantage of fewer lines and has therefore been chosen to demonstrate the field notes. The diagonal line is common to both triangles and so becomes the base line.

The survey is to be plotted to a scale of 1 : 500 so the degree of accuracy for field notes is 0.1 m. The field note pages shown in Figs. 2.35, 2.36 and 2.37 are in the sequence of booking. It will be noticed that line BC continues on the next sheet so the last entry is repeated on the next page.

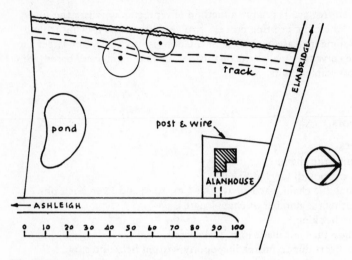

Fig. 2.33 Area to be surveyed

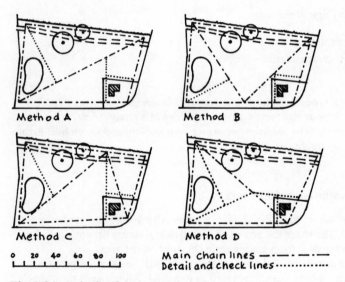

Main chain lines ‒·‒·‒·‒
Detail and check lines ············

Fig. 2.34 Methods of arranging chain lines

39

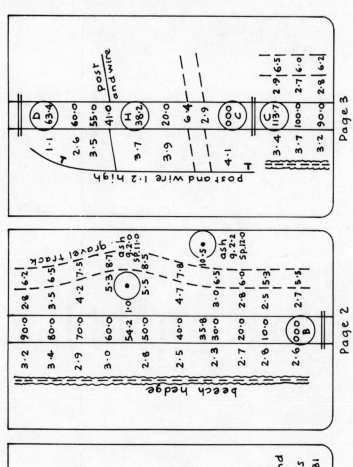

Page 3

Page 2

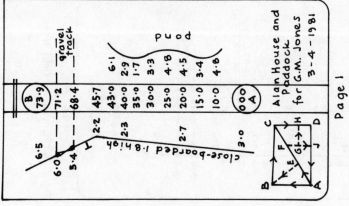

Page 1

Fig. 2.35 Chain survey field notes

40

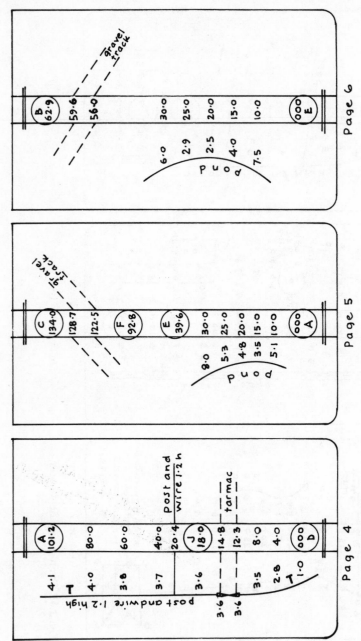

Fig. 2.36 Chain survey field notes

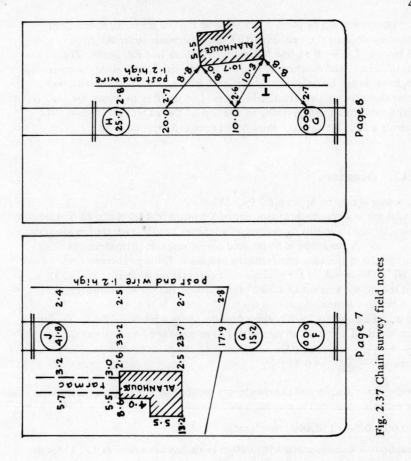

Fig. 2.37 Chain survey field notes

The survey can be plotted to a scale of 1 : 500 on an A3 sheet. Commence by drawing the base line AC. Using compasses strike AB from A and CB from C. Locate E on line AC and verify check line EB. Station D is struck from A and C, and checked by line FJ. If both check lines scale the correct length as shown on field notes the offsets can be plotted and detail drawn. Show north point and descriptive notes. For printing purposes the plan is traced with the north pointing to the top of the paper. To complete the drawing add a title panel to show scale, date and location of site.

2.12. Questions

Formulae shown in Appendix 1 (p. 154).
Qu 2.1 (a) A steel checking tape shows a tension of 44.5 N should be applied. Calculate the force this represents on a tension handle graduated in kilograms.

(b) A steel tape is to be used for setting-out permanent checking markers 20 m apart for standardizing purposes. The tape is correct when used at 20 °C. What will be the expansion or contraction when it is used at 32 °C, and should the amount be added or deducted from the measured length to find the true length?

Qu 2.2 *Example:* A 20 m chain is found to have stretched 0.08 m. Find the true length of a line which measures 110 m using the inaccurate chain.

$$\text{Error} = \frac{0.08 \times 110}{20} = 0.44 \text{ m.}$$

As the chain is stretched the true length will be more than that measured so the error is added to the original measurement.

$110.00 + 0.44 = 110.44$ m true length.

Question: A line measured with a 20 m chain has been recorded as 216.4 m but later it is found that the chain had been shortened in error and in fact measured 19.95 m. What is the true length of the line?

Qu 2.3 A clinometer measures a slope as $12° 30'$.
(a) Express as a gradient.
(b) Express as a percentage slope.
(c) If the slope length is 49.20 m what will be the horizontal length for plotting?

Qu 2.4 Figure 2.38 shows a pictorial view of an area with a chain line through it. AB is 114.5 m long. Show the field book entries for this line by estimating distances. The degree of accuracy for field measurements must be consistent with a plotting scale of 1 : 500.

Qu 2.5 The field plan shown in Fig. 2.39 is to be chain surveyed. On the plan show the most suitable arrangement of chain lines. The roads are not included in the survey.

Answers shown in Appendix 2 (p. 156)..

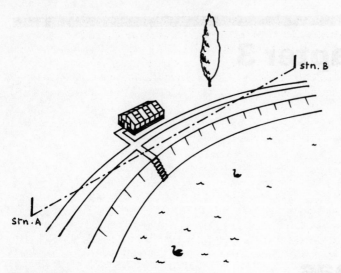

Fig. 2.38

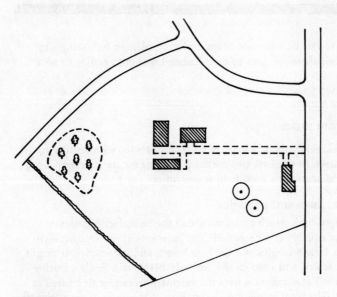

Fig. 2.39

Chapter 3

Areas

The area of a site can be obtained from a plotted plan or by taking site measurements that allow the area to be calculated without needing to plot the survey.

3.1. Areas from plans

The three methods explained below are suitable for sites with regular or irregular boundaries. In the first two methods the lengths are measured using a scale and the calculations are made in scaled units.

'Give and take' lines and triangles

Connecting straight lines which approximate to the boundary are drawn, so that any area lost equals the area gained. The junctions of the straight lines are connected to form triangles and the base length and perpendicular height of each triangle scaled and used in the formula BH/2. If a line separating two triangles is used as a common base the calculations can be simplified as $B/2 \times (H_1 + H_2)$. The total area is the summation of the areas of each triangle expressed in m^2 or hectares. See Fig. 3.1.

Average strip

The site is divided into strips of equal scaled width and the ends of each strip squared off using the 'give and take' principle. The length of each strip is scaled and the lengths totalled and multiplied by the strip width as in Fig. 3.2.

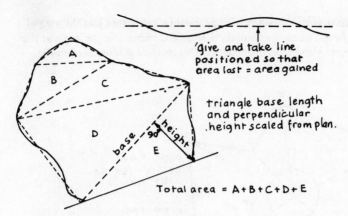

Fig. 3.1 Area by triangles with 'give and take' lines

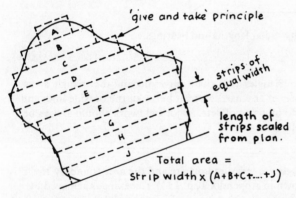

Fig. 3.2 Area by average strip

Planimeter

An instrument designed to find the area of any shape. Some planimeters are adjustable to suit the scale of the plan and give the answer directly in square metres. Others record the area full size to be multiplied by an appropriate factor to suit the scale of the drawing.

3.2. Areas direct from field measurements

Method A

This can be used if the boundary consists of straight lines. A theodolite, or a level with a graduated circle, is positioned fairly centrally on the site and radial lengths and bearings are measured to each change of direction of the boundary. The instrument is set to zero on one (reference object) point and all angles are then measured in a clockwise direction up to 360°.

The separate angles for each triangle must be calculated and the area of each triangle found using the formula ½ *ab* sine *C* where *C* is the angle at the centre with *a* and *b* the radiating lengths. This method is illustrated in Fig. 3.3.

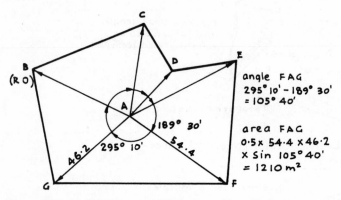

Fig. 3.3 Area of land by radial lengths and bearings

Method B

The site is divided into triangles as in chain surveying and the triangle areas are calculated. The areas of any narrow strips between the outside lines and the boundary are found by using offsets. Major fieldwork should always be checked, in this case the triangles.

Triangles

The sides of each triangle are measured on site and the area found by the 'S' formula (formulae shown in Appendix 1, p. 154). Checking is achieved by site measuring between opposite corners of adjacent triangles to form pairs of triangles which, when calculated, should give the same area as those first considered.

Narrow strips using offsets

Square offsets can be used to enable the area of a narrow strip of land to be calculated. Where the strips form the area between the outside survey lines and the boundary a small area of ground will be unaccounted for beyond the corner stations. This area will be small compared with the total area and can be estimated from the offsets taken.

(a) Separate trapezia. Used where the strip is bounded by fairly straight lines. Offsets from the survey line are taken at each change of direction of the boundary line (formula shown in Appendix 1, p. 154). Example shown in Fig. 3.4.

(b) Trapezoidal rule. Used where the boundary is irregular. It assumes straight lines connecting the offset ends. In practice the area lost will approximately

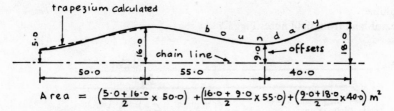

$$\text{Area} = \left(\frac{5\cdot0 + 16\cdot0}{2} \times 50\cdot0\right) + \left(\frac{16\cdot0 + 9\cdot0}{2} \times 55\cdot0\right) + \left(\frac{9\cdot0 + 18\cdot0}{2} \times 40\cdot0\right) \text{ m}^2$$

Fig. 3.4 Area by separate trapezia

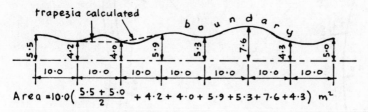

$$\text{Area} = 10\cdot0\left(\frac{5\cdot5 + 5\cdot0}{2} + 4\cdot2 + 4\cdot0 + 5\cdot9 + 5\cdot3 + 7\cdot6 + 4\cdot3\right) \text{ m}^2$$

Fig. 3.5 Area by trapezoidal rule

equal the area gained. Any number of offsets from the survey line are taken at regular intervals. It is unlikely that the length of line will be exactly a multiple of the offset interval. The area between the last regular offset and the offset at the end of the line is found by a separate trapezium (formula shown in Appendix 1, (p. 154). Example shown in Fig. 3.5.

(c) Simpson's rule. This is generally used only where the boundary consists of continuous curves but can in fact be applied to any shape. For curves it gives a more realistic result than the trapezoidal rule which, because of the straight lines connection to offset ends, would lose area continuously or gain it depending on the direction of the curve. Unlike the trapezoidal rule, Simpson's rule works only with an odd number of regularly spaced offsets giving an even number of strips. The area between the last odd offset and the offset at the end of the line should be treated as a separate trapezium (formula shown in Appendix 1, (p. 154). Example shown in Fig. 3.6.

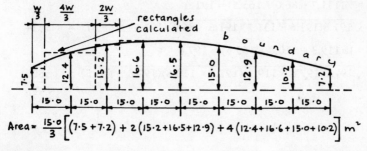

$$\text{Area} = \frac{15\cdot0}{3}\left[(7\cdot5 + 7\cdot2) + 2(15\cdot2 + 16\cdot5 + 12\cdot9) + 4(12\cdot4 + 16\cdot6 + 15\cdot0 + 10\cdot2)\right] \text{ m}^2$$

Fig. 3.6 Area by Simpson's rule

Example
Area of land from field notes Fig. 3.7

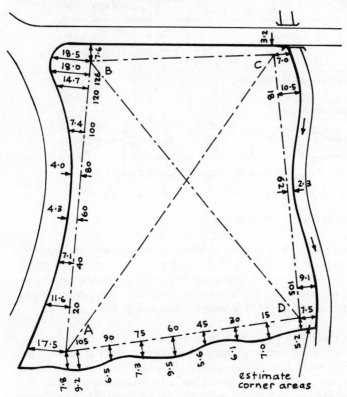

Line lengths AB 131·7, BC 86·0, CD 119·3, DA 110·8, AC 165·5, BD 152·1

Fig. 3.7 Area of land from field measurements

Triangles by 'S' formula

	Area m²

ABC S = ½(131.7 + 86.0 + 165.5) = 191.6

Area = $\sqrt{191.6(191.6 - 131.7)(191.6 - 86.0)(191.6 - 165.5)}$ = 5624

CDA S = ½(119.3 + 110.8 + 165.5) = 197.8

Area = $\sqrt{197.8(197.8 - 119.3)(197.8 - 110.8)(197.8 - 165.5)}$ = 6606

12 230

Check

ABD $S = \frac{1}{2}(131.7 + 152.1 + 110.8) = 197.3$

Area $= \sqrt{197.3(197.3 - 131.7)(197.3 - 152.1)(197.3 - 110.8)}$
$= 7114$

BCD $S = \frac{1}{2}(86.0 + 119.3 + 152.1) = 178.7$

Area $= \sqrt{178.7(178.7 - 86.0)(178.7 - 119.3)(178.7 - 152.1)}$
$= 5116 + 7114 = 12\ 230$ checks

Offsets AB by Simpson's rule 0 to 120

Area $= \frac{20}{3} [(17.5 + 14.7) + 2(7.1 + 4.0) + 4(11.6 + 4.3 + 7.4)]$ $\qquad = \qquad 984$

Offsets AB by trapezia 120 to 131.7

Area $= (\frac{14.7 + 18.0}{2} \times 6) + (\frac{18.0 + 18.5}{2} \times 5.7)$ $\qquad = \qquad 202$

Offsets BC by trapezia

Area $= \frac{7.6 + 3.2}{2} \times 86.0$ $\qquad = \qquad 464$

Offsets CD by trapezia

Area $= (\frac{7.0 + 10.5}{2} \times 18) + (\frac{10.5 + 2.3}{2} \times 44) +$

$(\frac{2.3 + 9.1}{2} \times 43) + (\frac{9.1 + 7.5}{2} \times 14.3)$ $\qquad = \qquad 803$

Offsets DA by trapezoidal rule 0 to 105

Area $= 15 \left[(\frac{5.2 + 9.2}{2}) + 7.0 + 6.1 + 5.6 + 9.5 + 7.3 + 6.5 \right]$ $\qquad = \qquad 738$

Offsets DA by trapezium 105 to 110.8

$\frac{9.2 + 7.8}{2} \times 5.8$ $\qquad = \qquad 49$

Estimated corner areas of A, B, C, D

$(20 \times 8) + (17 \times 7) + (7 \times 3) + (7.5 \times 5)$ $\qquad = \qquad \underline{338}$

$\qquad\qquad\qquad\qquad\qquad\qquad\qquad\qquad$ Total area $\quad = \quad 15\ 808\ \text{m}^2$
$\qquad\qquad\qquad\qquad\qquad\qquad\qquad\qquad\qquad\qquad\qquad = \quad 1.5808\ \text{ha}$

50

3.3. Errors in calculating areas

Gross errors
(a) Mis-scaling plotted plans.
(b) Misreading field measurements.
(c) Miscalculation.

Systematic errors
Planimeter out of adjustment.

Remedy

All areas must be checked as follows.
(a) Plotted plans. Use different method.
(b) Site measurements. Use different figures if available otherwise recalculate each stage.

3.4. Questions

Formulae shown in Appendix 1 (p. 154).
Qu 3.1 Find the area in hectares of the plot of land shown in Fig. 3.8 by:
(a) 'Give and take' lines and triangles.
(b) The average strip method.

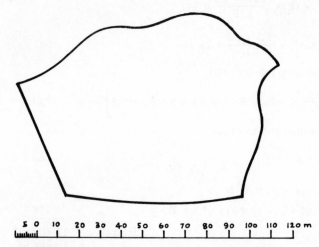

5 0 10   20 30 40 50 60 70 80 90 100 110 120 m

Fig. 3.8

Qu 3.2 Calculate in hectares the area between the road and the stream from station A to station B as shown on the chain survey field book (Fig. 3.9).
Qu 3.3 Figure 3.10 shows a measured field. Using the 'S' formula calculate the combined area of ABC and ACD in hectares. Check the area by adding ABD and BCD.

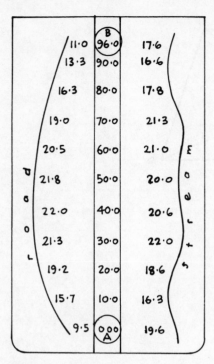

Fig. 3.9

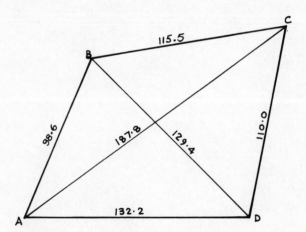

Fig. 3.10

Qu 3.4 Calculate the area of the building plot shown in Fig. 3.11 using the formula ½ *ab* sine *C*. Answer in hectares.
Answers shown in Appendix 2 (p. 156).

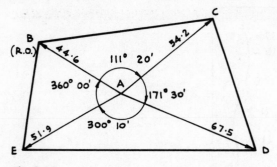

Fig. 3.11

Chapter 4

Levelling

Levelling is the term used for the process of finding the height of various points on the ground or parts of a structure above a certain horizontal reference surface.

Vertical measurements are taken by sighting horizontally through a mounted telescope onto a graduated staff held vertically. By giving the first point a certain height above the reference surface all the points can be related to this surface as illustrated in Fig. 4.1.

4.1. Instruments

Parts of the telescopic level

The instrument used for levelling is called a level and is basically a telescope with a bubble tube or similar levelling device attached so that a horizontal sight line is obtained. The instrument is mounted on a tripod for stability.

Telescope

Usually in the form of a tube, telescopes have a large object lens at one end and a small eyepiece at the other. Between these two is the image focusing lens which is moved backwards and forwards.

The range of magnification for the telescope varies from 10X to more than 40X. This number indicates the apparent increase in size of the object compared with the size viewed with the naked eye. The larger the magnification, the greater the length of sight possible. Some telescopes cause the image to appear inverted while others show an erect image.

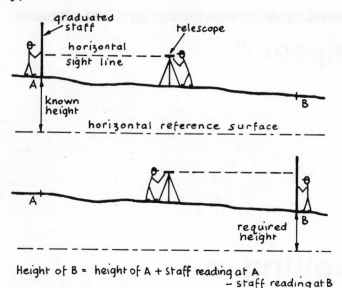

Height of B = height of A + staff reading at A
                              − staff reading at B

Fig. 4.1 Principle of levelling

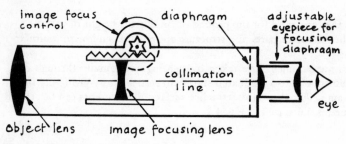

Fig. 4.2 Internally focusing telescope

Diaphragm

This carries the cross lines which are visible when sighting through the telescope. In early instruments hairs were used to form the lines and although the lines are now etched on glass graticules or reticules the term crosshairs is still used. The crosshairs are focused by screwing the eyepiece in or out. There are several different glass graticule markings used. Three of the more common ones are shown in Fig. 4.3.

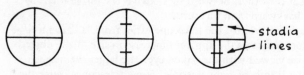

Fig. 4.3 Example of graticule markings

Collimation line

This is the axial line through the telescope passing centrally through the lenses and the diaphragm where it is shown by the central horizontal mark. For a level to be in correct adjustment the collimation line must be horizontal.

Stadia lines

The graticule vertical line normally carries two short horizontal lines equally spaced above and below the collimation line. The distance between these short lines appearing on the staff when multiplied by a constant will give the horizontal distance from the level to the staff. In modern instruments the constant is 100. This is a method of surveying called tacheometry where theodolites as well as levels are used.

Bubble tube

A curved glass tube is partially filled with a spirit leaving an air pocket which forms the bubble. Marks are etched on the tube to show when the bubble is central. Some high quality levels have a system of prisms which enable the surveyor to see both ends of the bubble side by side. When these are made to coincide the bubble is central. Such bubbles are called coincidence or split bubbles.

Classification of levels

Manufacturers produce levels for specific uses and models can be classed as builders', engineers', or precise instruments. This is because building sites are generally smaller than engineering sites and a higher degree of accuracy is demanded in bridge building, for example, than in normal building work.

Builders' levels

These have a low magnification but the quality of design and manufacture is adequate for normal building work.

Engineers' levels

These have a high magnification. A high quality instrument which is consequently more expensive than the builders' level.

Precise levels

Levels used for geodetic surveying requiring great accuracy over long distances. They have an optical micrometer for reading a staff faced with invar steel to minimize thermal movement.

Types of level

Within the three classes of level already described there are three basic types of instrument: dumpy, tilting and automatic levels. All have some form of horizontal locking clamp and slow-motion screw to direct the telescope centrally onto the staff. Most have a horizontal circle graduated in degrees so that horizontal angles can be set-out, but generally without the accuracy that can be achieved using a theodolite.

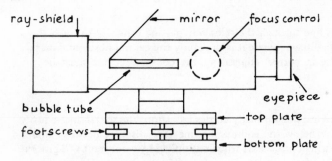

Fig. 4.4 Diagram of a dumpy level

Dumpy level

Although on most building sites all levels are loosely referred to as 'dumpies' the name, in fact, applies to one particular type of level — which has a telescope which is free to rotate at right-angles to the vertical axis. Assuming that the instrument is in perfect adjustment the collimation line will be horizontal whichever way the telescope is pointed. Figure 4.4 shows a diagram of a 'dumpy'. The instrument's bottom plate is screwed onto the tripod and the bubble tube levelled by the three footscrews as follows.

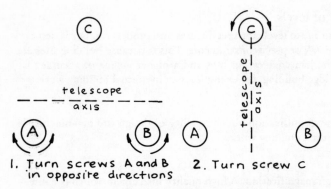

Fig. 4.5 Plan of footscrews showing method of centring bubble

1. Position the telescope so that the bubble is parallel with two of the footscrews.
2. Rotate the two footscrews at the same time in opposite directions, the surveyor's thumbs either moving towards or away from each other, until the bubble is central.
3. Turn the telescope 90° and rotate the third footscrew until the bubble is central.
4. Turn the telescope in all directions and the bubble should remain central. If the bubble does not stay central when the telescope is rotated 180° the instrument requires attention as described under 'permanent adjustments' later in this chapter.

The image seen through the dumpy level is inverted.

Tilting level

This instrument was developed from the dumpy level and is probably the most popular level used on construction sites. The difference between this and the dumpy level is that the telescope is not fixed at 90° to the vertical axis.

It has two spirit levels, a tubular one fixed to the telescope parallel to the collimation line, and a circular one fixed to the top plate. The term 'Quickset' is a trade name for a certain tilting level made by a particular manufacturer. The term is now used generally to describe any tilting level having a dome and cup base for quickly levelling the circular bubble. Engineering class levels centralize the circular bubble by three footscrews.

With all tilting levels the telescope bubble must be centred by turning the tilting screw when taking a staff reading. Figure 4.6 shows the parts of the tilting level. Depending on the particular model the image seen through the tilting level is erect or inverted.

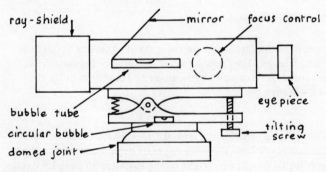

Fig. 4.6 Diagram of a tilting level

The advantage of the tilting level over the dumpy level is the ease with which it can be set up and the accuracy that can be achieved. Precise levels are invariably of the tilting type with coincidence bubbles.

Automatic level

This type of level once set up maintains a horizontal collimation line automatically without further adjustment. Another advantage is that, due to the optical design, it shows an erect image through the telescope.

To achieve a horizontal sight line automatically manufacturers use a variety of mechanisms which involve some form of pendulum prism device. Figure 4.7 shows diagrammatically how one system operates. To operate the instrument it needs only to be roughly levelled to centralize the circular bubble by use of dome and cup joint or three footscrews.

To verify that the self-levelling mechanism is operating freely the surveyor should tap the tripod while looking through the telescope: the image should appear to swing and return to its original position.

58

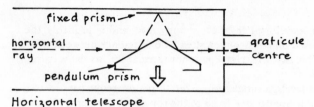

Horizontal telescope

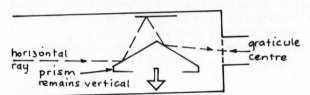

Tilted telescope

Fig. 4.7 Principle of Watts Autoset levels

The advantages of automatic levels are their speed of setting-up and use, and also their great accuracy. They are, however, unsuitable for use where there is vibration caused, for example, by wind on tall buildings or by the close proximity of vibrating plant.

Tripod

Wood or aluminium alloy is used for the legs of tripods which may be rigid or telescopic. The latter type have the advantages of being easier to transport and store, can be adjusted to suit the height of the surveyor and can be fixed at different lengths on sloping ground. Some tripod heads incorporate a circular bubble for levelling.

Levelling staff

Staves are made in wood, glass fibre or metal. They can be a telescopic, sectional or folding pattern in lengths of 3, 4 and 5 metres. Circular bubbles are sometimes incorporated for maintaining verticality. Figure 4.8 shows the 'E' type face marking specified in BS 4484:1969. The graduation marks are 10 mm deep with 10 mm spaces between the marks commencing from the bottom of the staff. Different colours, normally red and black, are used for graduation marks in alternate metres. Direct readings of the staff can be made to 0.01 m and estimated readings to 0.001 m.

Staff base plate

The purpose of this accessory, illustrated in Fig. 4.9, is to prevent the staff sinking into soft ground. This is particularly important at change points and it is often referred to as a change plate. It is made from steel in the shape of a triangle with each corner turned down and has a domed centre. A carrying chain is attached.

59

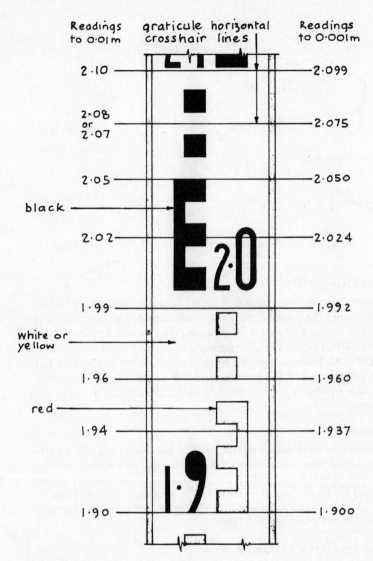

Fig. 4.8 Staff markings to BS 4484:1969 with example readings

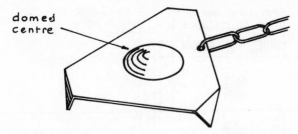

Fig. 4.9 Staff base plate

4.2. Levelling definitions

Datum

A reference surface to which levels are related. It can be given any height value.

Ordnance Datum (OD)

The most convenient datum for points on the land is mean sea level but as this varies around the coast of Britain it was established at one particular point. Ordnance datum of zero is the mean sea level recorded by the Ordnance Survey at Newlyn in Cornwall and established by taking hourly readings from 1915 to 1921. All levels on current OS maps are related to OD Newlyn but old maps were related to the now obsolete OD Liverpool. The difference between these two datums varies throughout the country due to errors in the original survey.

Local datum

Most construction work will be related to ordnance datum and this is particularly important when connecting to existing sewers, etc., as these will be related to OD on existing record drawings. For small isolated construction sites or small levelling operations it may not be necessary to relate levels to OD in which case a fixed point is made local datum and given an arbitrary height sufficient to avoid negative levels.

Ordnance bench-marks (OBM)

Various types of ordnance bench-marks have been established throughout Great Britain. They have been related to OD Newlyn by spirit levelling (telescope with bubble tube) and their heights are shown on large-scale OS maps and published bench-mark lists. The lists state the current bench-marks within a given area, their description, grid reference, altitude related to OD, height above ground, and year when last levelled.

There are four types of OS bench-marks.

Fundamental bench-marks

These are spaced about 40 km apart and used only by the Ordnance Survey.

Flush brackets

These are spaced at about 1.5 km intervals on OS levelling lines and on triangulation pillars. They consist of a slotted metal plate fixed to the vertical wall face. A shelf bracket clips to the plate to support the staff. A flush bracket bench-mark is shown on the triangulation pillar in Fig. 1.10.

Cut bench-marks

These are the common bench-marks spaced at approximately 300 m intervals in city centres and at 1000 m intervals in open rural areas. They are cut in vertical brick or stone faces. The height of the bench-mark refers to the centre of the cut horizontal bar above the broad arrow, shown in Fig. 4.10. A cut bench-mark (BM 114.98) is shown on a house in Fig. 1.10.

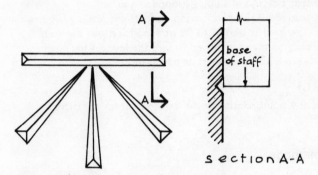

Fig. 4.10 Ordnance Survey cut bench-mark

Horizontal surface bench-marks

Three types are used, the Bolt BM, Rivet BM, and the Pivot BM which requires a 16 mm diameter ball-bearing to support the staff.

Temporary or transferred bench-marks (TBM)

It is normal practice to establish one or more reference points on site to which construction work can be related. Where possible this should be part of an existing structure such as a doorstep, inspection chamber cover or road kerb. A steel bolt grouted in position makes a positive point on which to stand the staff. Where no suitable feature is available a wooden peg or steel bar should be concreted into the ground. Further information on TBM construction is given in Chapter 7.

If the work being undertaken is related to ordnance datum the level of the TBM relative to OD must be established. Where a local datum applies the level of the TBM must be related to that datum. If local datum has not been established the TBM itself may be made the local datum.

Backsight (BS)

The first staff reading of a levelling operation is always called a backsight and the first reading after the level has been moved is also a backsight.

Foresight (FS)

The last staff reading of a levelling operation is always a foresight and the last reading before the level is moved is also a foresight.

Intermediate sight (IS)

Staff readings which are neither backsights nor foresights are intermediate sights.

Change point

A change point is the point at which the position of the level has to be changed. This may occur for any of the following reasons:
(a) staff being too far away for an accurate reading;
(b) staff cannot be seen because of sloping ground;
(c) staff cannot be seen due to an obstruction. A foresight reading is taken onto the staff. The level is moved to its new position but the staff remains on the same spot and is turned to face the level. A backsight reading is now taken with the level in its new position.

Reduced level (RL)

This is the height of any point relative to the datum used for the survey.

4.3. Levelling methods

Equal distance back and foresights

Referring to the principles of levelling outlined at the beginning of this chapter and illustrated in Fig. 4.1 it will be seen that the process relies on horizontal collimation lines. In practice it is extremely difficult to achieve absolute horizontality and the presence of some collimation error must always be assumed. If, on a certain instrument, the collimation line tilts slightly upwards from the horizontal, the angle of tilt will remain the same whatever direction the telescope is pointed. If all distances from the level to the staff are the same the collimation errors will be equal and will cancel out so the correct difference in ground level will still be established. This is shown in Fig. 4.11.

Because of collimation error the backsight and foresight taken from each set-up of the instrument should be approximately at equal distances from the instrument. Using this sytem errors do not accumulate. Intermediate sights cannot normally be kept to any particular length so slight errors are inevitable depending on the accuracy of the instrument and the length of sight.

Reading the staff

The maximum length of sight depends on the magnification and quality of the instrument and the fineness of staff reading required. Staff readings for typical instruments used on small construction sites should be limited to a distance of approximately 60 m.

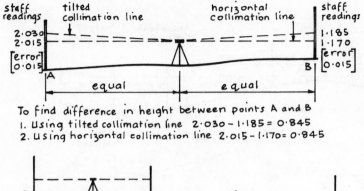

To find difference in height between points A and B
1. Using tilted collimation line 2.030 - 1.185 = 0.845
2. Using horizontal collimation line 2.015 - 1.170 = 0.845

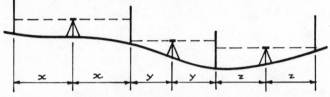

Fig. 4.11 Equal distance back and foresights

The fineness of staff readings depends on the purpose of the levelling. Backsights and foresights should normally be read to 3 decimal places as should setting-out work generally. Intermediate sights on rough ground and for setting-out excavation depths are adequate to 2 decimal places.

Examples of staff readings are shown in Fig. 4.8.

Booking levels

Where possible the first reading is taken on to an OBM or TBM. In the absence of an established bench-mark the first reduced level is given an arbitrary rounded level (e.g., 50.00) sufficient to avoid negative RLs. It is possible to relate the readings to a reference point part-way through or at the end of a levelling operation and to calculate the previous reduced levels by working back from the known reduced level.

Figure 4.12 shows staff readings taken at 20 m intervals on an existing road. Table 4.1 shows how the staff readings are entered on the page of a level book. The staff is held on the OBM and the first reading of 1.944 is entered as a backsight. Staff readings at chainage 0 and 20 are entered as intermediate sights. Because of the sloping ground the staff at chainage 60 would be out of sight so chainage 40 is made a change point. A foresight is taken with the staff at chainage 40 and the staff remains at that point. The level is moved to the second position and a backsight is taken at chainage 40 and entered on the same line as the previous reading. Intermediate readings are taken at chainage 60 and 80 and a foresight reading of 0.775 taken at chainage 100 being the last road level required. To verify the accuracy of the work a check back to the OBM is taken with the instrument at the third position. A backsight of 2.114 on chainage 100 and a foresight of 0.816 on the OBM complete the staff readings.

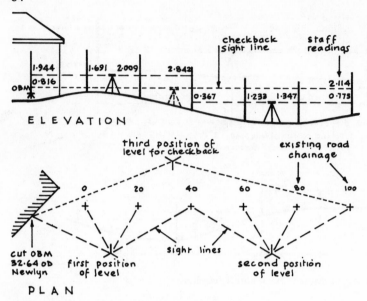

E L E V A T I O N

P L A N

Fig. 4.12 Staff readings on existing road

Table 4.1 Level book entries of staff readings shown in Fig. 4.12

BS	IS	FS	RL	Remarks
1.944			32.640	OBM
	1.691			Chainage 0
	2.009			Chainage 20
0.367		2.842		Chainage 40
	1.233			Chainage 60
	1.347			Chainage 80
2.114		0.775		Chainage 100
		0.816		OBM

Where staff readings fill a page of a notebook the calculations on each page must be checked independently. Where the last staff readings on a page are for a change point only the foresight should be entered. The backsight should be used to start the next page. Where the last staff reading on a page is an intermediate sight it should be entered as a foresight and the same reading entered as a backsight at the beginning of the following page.

Reducing levels
Reducing levels is the process of calculating the required reduced levels from staff readings by either the 'rise and fall' or the 'height of collimation' method.

Rise and fall method

This method is illustrated in Table 4.2 and uses the staff readings from Fig. 4.12 entered in Table 4.1.

Table 4.2 Levels reduced by rise and fall method

BS	IS	FS	Rise	Fall	RL	Remarks
1.944			—	—	32.640	OBM
	1.691		0.253		32.893	Chainage 0
	2.009			0.318	32.575	Chainage 20
0.367		2.842		0.833	31.742	Chainage 40
	1.233			0.866	30.876	Chainage 60
	1.347			0.114	30.762	Chainage 80
2.114		0.775	0.572		31.334	Chainage 100
		0.816	1.298		32.632	OBM
4.425		4.433	2.123	2.131	32.640	
		4.425		2.123	32.632	
		0.008		0.008	0.008	

1. Enter staff readings in appropriate columns.
2. Total the backsights and total the foresights, subtracting the smaller from the larger to find the difference.
3. Blank out the first spaces in both the rise and fall columns. Subtract each successive staff reading from the previous staff reading and enter in the 'rise' or 'fall' columns. Figure 4.13 shows the sequence of staff readings to be subtracted.

   It will be seen from the staff readings whether the ground has risen or fallen between the two points. If the second reading is larger than the

This applies to the staff readings entered in Table 4·1 and shown reduced in Table 4·2

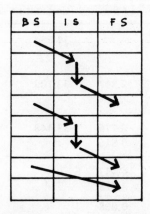

Fig. 4.13 Example of subtraction sequence for rise and fall method

first the ground level has fallen. If the second reading is smaller than the first the ground level has risen. When using a calculator subtract each successive staff reading from the previous one and if the answer is positive it is a rise, if negative it is a fall.

1.944 − 1.691 = 0.253 (rise)
1.691 − 2.009 = −0.318 (fall)
2.009 − 2.842 = −0.833 (fall)

Note: Do not show minus signs in the 'fall' column.

4. Total the 'rise' column and total the 'fall' column, subtract the smaller total from the larger to find the difference. This should be the same as the difference between the total backsights and foresights.
5. The first RL should next be entered. If this is not a known value from an OBM or TBM, show a round figure such as 10.000, large enough to avoid negative results.
6. Calculate the other RLs by adding rises to the previous RL and deducting falls from the previous RL in turn.

32.640 RL, OBM
32.640 + 0.253 = 32.893 RL chainage 0
32.893 − 0.318 = 32.575 RL chainage 20

7. To check that RLs have been correctly calculated take the first and last RLs and subtract the smaller from the larger to find the difference. This figure should agree with the two previous differences found.

Height of collimation method

This method is also known as the 'height of plane of collimation' and the 'height of instrument' method. Table 4.3 illustrates this method using the staff readings from Fig. 4.12 entered in Table 4.1.

Table 4.3 Levels reduced by height of collimation method

BS	IS	FS	Collimation	RL	Remarks
1.944			34.584	32.640	OBM
	1.691			32.893	Chainage 0
	2.009			32.575	Chainage 20
0.367		2.842	32.109	31.742	Chainage 40
	1.233			30.876	Chainage 60
	1.347			30.762	Chainage 80
2.114		0.775	33.448	31.334	Chainage 100
		0.816		32.632	OBM
4.425		4.433		32.640	
		4.425		32.632	
		0.008		0.008	

This applies to the
staff readings entered
in Table 4·1 and shown
reduced in Table 4·3

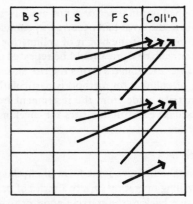

Fig. 4.14 Example of subtraction sequence for height of collimation method

1. Enter staff readings in appropriate columns.
2. Total the backsights and total the foresights, subtract the smaller from the larger to find the difference.
3. To find the height of collimation always add to the RL the BS taken at that point.

$$32.640 + 1.944 = 34.584 \text{ collimation}$$

4. Reduced levels are found by subtracting from the height of collimation all the staff readings taken while the instrument was in that position. Fig. 4.14 shows the sequence of staff readings to be subtracted.

$$34.584 - 1.691 = 32.893 \text{ RL chainage } 0$$
$$34.584 - 2.009 = 32.575 \text{ RL chainage } 20$$
$$34.584 - 2.842 = 31.742 \text{ RL chainage } 40$$

Chainage 40 is a change point where the instrument was moved to a new position giving it a new height of collimation.

New height of collimation = RL + BS
= 31.742 + 0.367 = 32.109 collimation.

5. Take the first and last RLs and by subtracting the smaller from the larger find the difference. This should be the same difference as found when totalling backsights and foresights. On examination it will be seen that the RL check only verifies the backsight and foresight readings. If mistakes were made when calculating RLs from intermediate sights it would not alter the first and last RLs.

Comparison of methods

Although the height of collimation method involves less figure work and is thus quicker, it suffers from the great disadvantage that the intermediate sights are not checked and for some levelling operations most levels taken are intermediates. There is a complicated mathematical check that theoretically could be applied but in practice is not used. It would detect an error but not pinpoint where the error had occurred so complete recalculation would be required. In practice if the height of collimation method is used each calculation for intermediate sights should be repeated to minimize the possibility of errors. Because it eliminates errors the rise and fall method is recommended. The height of collimation method is particularly convenient when several staff readings are made from the instrument in the same position and is generally used on construction sites for checking the height of various parts of the structure.

Checkback

All surveying should have some form of check applied to detect errors. At the end of a levelling operation on site a series of backsights and foresights should be taken either back to the starting point or to another known site level if closer. The length of sights can be longer than those taken in the first place. When returning to the starting point it is good practice to use the same change points as in the original levelling so that errors can be isolated.

Accuracy in levelling

On completion of checkback the final RL is likely to show a slightly different value from the known level and this difference is the closing error.

BS 5606:1978 *Code of Practice for accuracy in building* states permissible deviations in setting-out. The table below has been prepared from that document.

Instrument	Single sight up to 60 m	Per km
Builders' level	± 5 mm	
Engineer's level	± 3 mm	± 10 mm

The following formula is considered acceptable and is within the limits stated above.

(a) Dumpy and builders' tilting level
Permissible error = $\pm 20\sqrt{k}$ mm

(b) Automatic and engineers' level
Permissible error = $\pm 10\sqrt{k}$ mm
where k is total circuit in kilometres.

If the closing error is not within the limits of the permissible error the work should be re-levelled.

Example calculation. For the levelling undertaken in Fig. 4.12 the length of the circuit was approximately 120 m x 2 = 0.24 km. If permissible error is $20\sqrt{k}$ mm

$20\sqrt{0.24}$ mm = 9.8 mm

The RL error was

32.640 – 32.632 = 0.008 = 8 mm

and so is within permissible limits.

Allocation of closing error

The closing error should be distributed as equally as possible between each position of the instrument.

8 mm in 3 instrument positions = 3 mm + 3 mm + 2 mm.

The adjustment is made to all intermediate and foresights but not backsights. The distribution is cumulative.

First instrument position. Allocation 3 mm
Second instrument position. Allocation 3 + 3 = 6 mm
Third instrument position. Allocation 3 + 3 + 2 = 8 mm

Table 4.4 shows allocation of error for reduced levels shown in Tables 4.2 and 4.3.

Table 4.4 Allocation of closing error shown in Tables 4.2 and 4.3

Calculated RL	Correction	Corrected RL
32.640	+0.000	32.640
32.893	+0.003	32.896
32.575	+0.003	32.578
31.742	+0.003	31.745
30.876	+0.006	30.882
30.762	+0.006	30.768
31.334	+0.006	31.340
32.632	+0.008	32.640
32.640		32.640
32.632		32.640
0.008		0.000

Where flying levels are used to establish a TBM from a known reference level the checkback provides a closing error.

If this error is within the limits of the permissible error half of it is allocated to the TBM.

4.4. Types of levelling

Flying levels

This is the process used in transferring levels from one point to another. It consists only of backsights and foresights with the length of the sights being as long as possible to suit the instrument being used. Flying levels are used, for example, in setting-up a site TBM from an OBM and for checkback at the end of a levelling operation. Figure 4.11 (illustrating equal distance back and foresights) is an example of flying levels.

Grid levels

A grid of levels may be required for:

(a) contouring an area;
(b) calculating the volume of earthworks;
(c) finding ground levels for columns in a framed building;
(d) general appreciation of a site for development.

The square or rectangular grid is set-out on the ground with pegs using one of the methods described in Chapter 2 for forming right-angles. Alternatively the horizontal circle on the level can be used for 90° angles. The grid lines are lettered alphabetically in one direction and the lines at right-angles numbered consecutively. The level is positioned to take as many staff readings as possible and the height of collimation method of reducing levels is the more convenient one to use. To save time the staff readings are taken following a snaking route as A1, A2, A3, A4, B4, B3, B2, B1, C1, C2, etc. shown in Fig. 4.15.

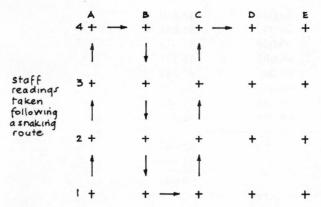

Fig. 4.15 Grid levels

Series or line levels

Where, for example, a drain or road is to be constructed, existing ground levels are required on the centre-line of the proposed work. The levels will be taken at regular intervals and also at change of gradient, each point being referred to by its chainage (running distance) from the commencement of the work. Change-points can be on a centre-line peg or any other suitable point. The levels in Fig. 4.12 are series or line levels. When levels have been reduced a longitudinal section can be drawn showing existing ground levels, proposed road levels, drain invert levels, etc., the invert of a drain being the inside bottom of a pipe or channel.

Cross-section levels

In road construction the earthwork involved in 'cutting' to remove soil and 'filling' to raise the ground level is calculated by preparing cross-sectional drawings at regular intervals along the route of the road. All cross-sections are known by their chainage from the beginning of the road. With the surveyor standing on the road centre-line, having his back to the chainage zero, levels are described as on his right or left. At each cross-section pegs are set-out at right-angles to the centre-line pegs. The pocket optical square or prism is the most useful instrument for setting-out the right-angles.

Levels are taken at regular intervals on the road centre-line and at regular intervals right and left of the centre peg at each cross-section. Alternatively cross-section levels can be taken at the ends of the cross-section and any change of gradient in-between.

Both longitudinal section levelling and cross-section levelling can be undertaken in one operation but frequent checkbacks should be made. It is good policy to establish TBMs along the course of a proposed road to form established reference points before section levelling commences. Figure 4.16 shows the plan of cross-section level points and illustrates the sequence and description of field booking.

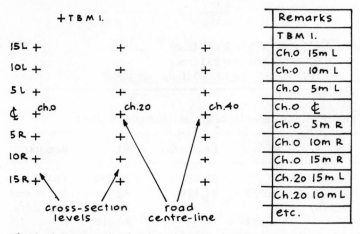

Fig. 4.16 Cross-section levels and booking method

Inverse levels

Inverse levels are taken with the staff in an inverted position to find the level of any feature above the collimation line and have no connection with a drain invert. Inverse levels are used, for example, to find the level of a bridge or beam soffit or used to continue levelling beyond a tall wall or similar obstacle. In each case the staff is turned upside-down and the base is held on the soffit or at the top of the wall. The staff reading is booked as a minus figure and calculated algebraically. Figure 4.17 shows examples of inverted staff readings with the methods of reducing the levels shown in Tables 4.5 and 4.6.

72

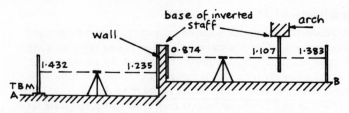

Fig. 4.17 Inverse staff readings

Table 4.5 Inverse levels reduced by rise and fall method

BS	IS	FS	Rise	Fall	RL	Remarks
1.432			—	—	30.000	TBM A
-0.874		-1.235	2.667		32.667	Top of wall
	-1.107		0.233		32.900	Arch soffit
		1.383		2.490	30.410	B
0.558		0.148	2.900		30.410	
0.148				2.490	30.000	
0.410			0.410		0.410	

Calculations   1.432 − (−1.235) = 2.667 (rise)
              −0.874 − (−1.107) = 0.233 (rise)
              −1.107 − 1.383   = −2.490 (fall)

Table 4.6 Inverse levels reduced by height of collimation method

BS	IS	FS	Collimation	RL	Remarks
1.432			31.432	30.000	TBM A
-0.874		-1.235	31.793	32.667	Top of wall
	-1.107			32.900	Arch soffit
		1.383		30.410	B
0.558		0.148		30.410	
0.148				30.000	
0.410				0.410	

Calculations 31.432 − (−1.235) = 32.667 RL
             32.667 + (−0.874) = 31.793 collimation
             31.793 − (−1.107) = 32.900 RL

4.5. Fieldwork procedures

Before visiting the site the surveyor should ascertain the location of the nearest OBM and site TBMs. Once he has arrived on site, pegs need to be set out to mark the levelling lines or grid.

Setting-up the instrument

The level should be positioned to take as many readings as possible from that point to minimize the number of times it has to be moved. Where possible avoid setting-up on public roads, site roads and pavements because of danger to traffic and pedestrians.

The tripod

The tripod head should be at a convenient height for the surveyor's use. It should be approximately levelled using telescopic legs. On a hill tripods with rigid legs should have one leg pointing up the slope. The legs should be about a metre apart and pressed firmly into the ground. Where provided, wingnuts should be tightened for use and loosened again when folding up the tripod.

Fixing the level

When removing the instrument from the box note how it should be replaced and close the box to keep out dust and leaves. For fixing the level to the tripod hold the telescope in the left hand and screw the bottom plate with the right hand. Where the tripod head forms the dome of a dome-and-cup fitting the instrument is screwed in position from below and the circular bubble centred at the same time. If footscrews are fitted they are used at this stage to centralize plate bubbles.

Focusing

The graticule crosshairs are focused to suit the vision of the surveyor by turning the eyepiece in or out until the crosshairs appear sharp. A piece of paper or the open levelling book held in front and slightly away from the telescope helps in this operation. The crosshairs will need no further adjustment unless used by another surveyor.

The telescope is directed to the staff using the top sights and is clamped in position. The surveyor looks through the telescope and centres onto the staff with the slow-motion screw. Turning the focusing knob on the side of the telescope causes the staff graduations to appear distinct. At this stage the tilting level bubble must be centred using the tilting screw.

Parallax

For correct focusing the staff image should fall exactly on the diaphragm and so should the eyepiece focal point. If either of these fail to happen, parallax is present and accurate staff readings cannot be made. To check for parallax move the eye up and down while viewing the staff through the telescope. If the crosshairs appear to float up and down relative to the staff graduations, parallax is present. It is removed by first checking that the crosshairs are correctly focused and sharp and then refocusing the telescope onto the staff.

Once the crosshairs appear to be fixed onto the staff while the eye is moved up and down, parallax has been eliminated and the staff reading can be taken.

Taking readings

Entries in the level book should be made in pencil and mistakes erased. The notes column should be filled in at the same time as staff readings are entered. To reduce errors the staff should be read, the reading booked, and the staff re-read as a check before directing the staffman to move on. The tripod should not be touched while readings are being taken. When the level is moved the tripod legs should be closed and the tripod carried in the upright position resting against the shoulder. When the level is set up in its new position all bubbles must be re-centred.

The external lens surface should be cleaned with soft linen or cotton rag moistened in water or alcohol.

Using the staff

Use the staff in its fully extended form only when readings demand it. When extending telescopic staffs ensure that each section is correctly located in position. This can be verified by looking at the graduations at the joint. When taking readings at a cut bench-mark hold the bottom of the staff centrally on the horizontal mark. For other readings the staffman stands behind the staff holding it by the sides so as not to obscure the staff face.

Staves must be held vertically for an accurate reading to be taken and this is easily maintained if a circular bubble is fitted. For other staves verticality is achieved by using the following method. The surveyor can see if the staff is leaning sideways and can signal for it to be held vertically by leaning his forearm to the vertical position, indicating which way the top of the staff must be moved. The surveyor cannot see if the staff is leaning towards or away from him so when the reading is being taken the staff holder slowly rocks the top of the staff towards the surveyor and away again. The surveyor takes the smallest reading, as this occurs when the staff is in the vertical position (see Fig. 4.18). Rocking the staff is not necessary on rough ground as a high degree of accuracy is not required.

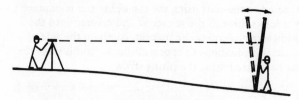

Fig. 4.18 Rocking staff for vertical reading

Change points should be carefully selected as subsequent levels depend on the accuracy of readings taken at a change point. They may have to be re-located so should be in a positive position with a solid surface such as the corner of an inspection cover, a road kerb or peg top. On soft ground a staff base plate, illustrated in Fig. 4.9, should be used.

Equal distance back and foresights are achieved by the staffman counting an equal number of paces before positioning the staff.

The staff should not be leaned against a wall as it is easily blown down and damaged. It should not be casually carried in the extended position over the shoulder. Care should be taken not to scratch the face of the staff which should be cleaned occasionally with a soapy rag.

Weather

Wind blowing on the level causes it to vibrate and at times makes staff reading impossible. Wind also makes the staff difficult to hold still.

Position the level to look away from the sun rather than towards it. The sun's rays can often be shielded from the object lens by holding the level book above the end of the telescope. The sun also causes the ground to shimmer, making reading of the lower part of the staff impossible.

Rain on the lens makes reading difficult and writing in the level book becomes a problem. When working in the rain a surveyor's umbrella may be used to protect the level. If the level becomes wet it should be dried with a soft cloth in a warm place. Never put a level away in its closed box if still wet as moisture may form on the lenses inside the telescope.

4.6. Adjustments to level

Temporary adjustments
These are the adjustments carried out every time the level is set-up.

1. Centralize bubble.
2. Focus crosshairs.
3. Focus image.
4. Check for parallax.

Permanent adjustments
These adjustments require the use of a spanner, tommy-bar, etc. All levels should be checked at regular intervals to find if such adjustments are necessary.

(a) To ensure that the bubble tube is at right-angles to the vertical axis
Note: This adjustment applies only to the dumpy level. Refer to Fig. 4.19(a) for illustration of stages.

1. The dumpy level is set-up on the tripod and the bubble tube centred with the footscrews.
2. The telescope is rotated horizontally 180°.
3. If the bubble has run off centre **half** the error must be removed with the bubble end-screws. See Fig. 4.19(b).
4. The bubble is then centred by the footscrews and should remain central in whatever direction it is pointed.

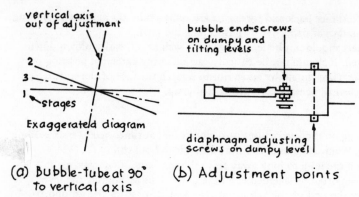

vertical axis
out of adjustment

2
3
1
stages

Exaggerated diagram

bubble end-screws
on dumpy and
tilting levels

diaphragm adjusting
screws on dumpy level

(a) Bubble-tube at 90°
to vertical axis

(b) Adjustment points

Fig. 4.19 Permanent adjustments to level

(b) To ensure that the collimation line is truly horizontal. This is called the 2-peg test
Note: This adjustment applies to all levels. Refer to Fig. 4.20 for illustration of stages.

Two-peg test: stage 1. Drive in two pegs 50 m apart and set-up the level exactly midway between them. Take staff reading on top of peg A and on top of peg B. As explained in the section on equal distance back and foresights (illustrated in Fig. 4.11) the readings, when subtracted, will give the true difference in height between the pegs.

Two-peg test: stage 2. Move the level beyond peg B and take staff readings on pegs A and B. If the level has no collimation error the difference between the readings will be identical to that found in stage 1. If the difference is not as previously found collimation error exists and, with the level remaining in the same position, adjustments should be made as described.

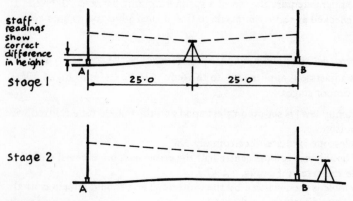

staff
readings
show
correct
difference
in height

A   B
Stage 1   25·0   25·0

Stage 2   A   B

Fig. 4.20 Two-peg test

Dumpy level. With the telescope bubble centred adjust the diaphragm top and bottom screws (Fig. 4.19(b)) until staff readings at A and B show the stage 1 difference.

Tilting level. Turn the tilting screw until staff readings at A and B show the stage 1 difference. Adjust the bubble end-screws (Fig. 4.19(b)) to centralize the bubble.

Automatic level. The usual adjustment is to the diaphragm as in the dumpy level but reference should be made to the manufacturer's handbook.

4.7. Errors

Gross errors

(a) Level incorrectly adjusted.
(b) Staff not properly extended.
(c) Staff not held vertically.
(d) Staff moved at change point.
(e) Misreading against stadia lines instead of horizontal crosshair.
(f) Misreading when staff appears inverted.
(g) Misreading due to closeness of staff so metre figures are not visible.
(h) Booking in wrong column.
(i) Booking different figures from those read.
(j) Reducing errors particularly intermediate readings using height of collimation method.

Remedy

Pay attention to points explained in this chapter, particularly:

(a) checking staff readings before and after booking;
(b) using frequent checkbacks;
(c) using rise and fall method wherever convenient;
(d) repeating height of collimation intermediate reading calculations.

Systematic errors

Caused by instrument being out of permanent adjustment.

Remedy

Where a surveyor has sole use of level it should be tested at regular intervals. All other levels should be tested before use.

4.8. Questions

Qu 4.1 Figure 4.21 shows where the telescope horizontal line cuts the staff for readings taken at one set-up of the level. Book these readings to 2 decimal

78

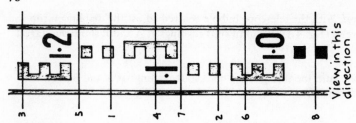

Fig. 4.21

places in numerical order and reduce by the rise and fall method. The RL of the first reading is 30.00.

Qu 4.2 Figure 4.22 shows a plan of rectangular grid levels taken on rough ground. The level was first set-up at W and later moved to Z. The staff readings in sequence were: TBM 0.83, A1 0.97, A2 1.04, A3 1.17, B3 2.25, B2 2.13, B1 2.10, B1 1.82, C1 2.69, C2 2.73, C3 2.86, TBM 0.56. The TBM RL is 20.79.

Book the readings and reduce by the height of collimation method.

```
      A            B            C
  3 +           +            +



  2 +      ⅄w    +     ⅄z     +
 +
TBM

  1 +           +            +
```

Fig. 4.22

Table 4.7 Question 4.3

BS	IS	FS	Rise	Fall	RL	Remarks
1.343					36.470	OBM
0.487		1.076				
1.007		2.103				
1.867		1.942				
0.952		0.875				TBM
1.763		1.486				
2.085		1.329				
1.848		1.741				
		0.788				OBM

Qu 4.3 A TBM was established on site from an OBM 400 m away and the checkback followed the same route but with different change points. Reduce the staff readings shown in Table 4.7 by the rise and fall method. If the closing error is within the permissible limit for a builder's tilting level show the adjusted reduced level of the TBM.

Qu 4.4 Figure 4.23 shows the elevation of staff and readings taken inside a building under construction. Book these readings and reduce by the height of collimation method.

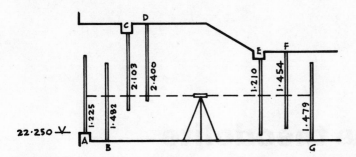

Fig. 4.23

Answers shown in Appendix 2 (p. 156).

Chapter 5

The theodolite

Simple angular measurements can be made with various surveying instruments but where accuracy is required the theodolite is used. It will measure horizontal and vertical angles to various degrees of accuracy depending on the quality of the instrument.

5.1. Classification of instruments

Manufacturers describe instruments as reading to a certain fineness of measurement. A direct reading is where the arrow or index mark coincides with a graduation line. An estimated reading is where the arrow or index mark falls between the graduation lines.

Builders' theodolites

Robust construction and ease of use are features associated with these instruments. They usually show divisions enabling direct angular measurements to one minute interval to be read. On construction sites the theodolite is invariably used for setting-out horizontal angles and a few economy instruments are made without the facility for measuring vertical angles.

Engineers' theodolites

These have higher magnification than the builders' instrument and will read to 20 or 10 seconds without estimation. Accessories available include a magnetic compass and $90°$ eyepiece enabling the telescope to point vertically for optical plumbing of tall structures.

Precision and one-second theodolites

High magnification and high accuracy are the qualities connected with these instruments which read directly to 1 second or less. They are mostly used for extensive land surveys demanding very fine limits of angular measurement.

5.2. Constructional principles of the theodolite

Although there are different types of theodolite certain basic features are common to them all. These are illustrated in Fig. 5.1.

Provision is always made for the following operations.

Plumbing

A vertical line under the centre of the instrument is found by (a) a suspended plumb-bob, (b) an adjustable plumbing rod with circular bubble or (c) an optical plummet providing a vertical sight line. Some instruments with optical plummets also carry a plumb-bob for initial rough centring.

Centring

It is difficult to set-up the instrument directly over a point by movement of the tripod alone so provision is made for the instrument to slide bodily once on the tripod. This is achieved by a centring facility on the tripod head or on the instrument itself.

Levelling

The plate bubble is mounted on the upper plate (also called an alidade) and is centred by the 3 footscrews. When vertical angles are being measured the altitude bubble must be centred to provide a horizontal index. Some instruments have automatic horizontal indexing thus dispensing with the altitude bubble.

Horizontal angles

The horizontal circle is divided into 360 degrees and parts of a degree in a clockwise direction. It is free to rotate and can be clamped to the tribrach leaving the upper plate free to rotate, or it can be clamped to, and rotate with, the upper plate.

Both upper and lower plates can be clamped together. The clamps are provided with slow-motion screws.

Vertical angles

The vertical circle is divided in the same way as the horizontal circle. It is permanently fixed to the telescope. The origin when numbering the degrees varies with manufacturer.

If the telescope is clamped in the horizontal position the instrument can be used as a level.

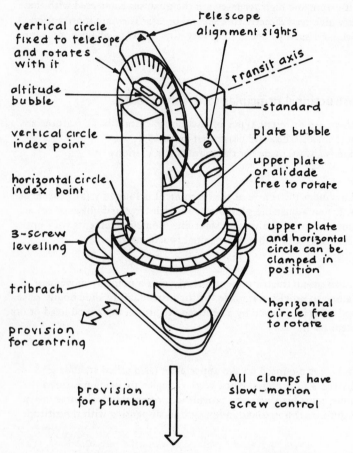

telescope can be clamped in any position

telescope
alignment sights

vertical circle
fixed to telesope
and rotates
with it

transit axis

altitude
bubble

standard

vertical circle
index point

plate bubble

upper plate
or alidade
free to rotate

horizontal circle
index point

3-screw
levelling

upper plate
and horizontal
circle can be
clamped in
position

tribrach

horizontal
circle free
to rotate

provision
for centring

provision
for plumbing

All clamps have
slow-motion
screw control

Fig. 5.1 Diagram showing essential features of theodolite

Sighting

The telescope is of the same basic construction as that used on the levels shown in Fig. 4.2. Various graticule markings are used, all of which have horizontal and vertical crosshairs. The image may be erect or inverted depending on the design. The telescope rotates vertically through 360° and can be clamped in any position and adjusted by a slow-motion screw.

External alignment sights are fixed to the top and sometimes also the bottom of the telescope.

5.3. Types of instrument

Theodolites have either vernier or optical reading systems.

Vernier theodolite

Although these are now considered obsolete and have mostly been replaced with more modern instruments many continue to be used. One is shown in Fig. 5.2.

Fig. 5.2 Vernier theodolite

Plumbing

A plumb-bob is suspended centrally below the instrument.

Centring

Provision is made for the instrument to slide on the tribrach and is clamped from below.

Horizontal circle

The horizontal circle is usually marked in degrees from 0° to 360° with 20 minute divisions.

Upper plate

The upper plate carries two vernier scales 180° apart for reading the horizontal circle.

Vertical circle

The vertical circle usually shows degrees and 20 minute divisions with each quadrant numbered to 90°, the zeros being coincident with the telescope sight axis.

Altitude bubble

The altitude bubble is fixed to a plate carrying two vernier scales which are positioned on a horizontal line to read opposite sides of the vertical circle. The bubble is centred by slow-motion screws bearing against one of the standards.

Telescope

The telescope shows an inverted image and is focused by a knob at the end of the transit bearing.

Vernier reading system

These scales enable small divisions to be evenly subdivided. The notes shown below refer to Fig. 5.3 which illustrates a vernier scale used to sub-divide main scale divisions into 10 equal parts.

Construction. The vernier scale is always $n - 1$ main scale divisions in length. In this case $n = 10$ so the vernier scale is 9 main scale divisions long.

Reading. It will be seen that each vernier division is 0.1 short of a main scale division. If the vernier scale in Fig. 5.3(a) were to be moved until the 1 mark coincided with a main scale mark the index arrow would have moved 0.1 of a main scale division. If the vernier scale 2 mark coincides with a main scale mark the arrow has moved 0.2 and so on. Apart from 0 and 10 only one mark at a time on the vernier scale will coincide with a mark on the main scale.

When reading a vernier scale always record the complete divisions shown by the index arrow first and then add the vernier coincidence reading to it. Fig. 5.3(b) shows a random reading.

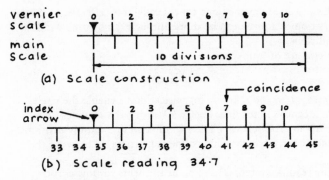

Fig. 5.3 Vernier scale principle

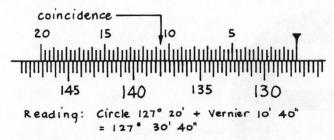

Fig. 5.4 Theodolite horizontal circle vernier

Illustrated in Fig. 5.4 is a theodolite horizontal circle with degrees and 20 minute divisions. The vernier scale shows minutes and 20 second divisions.

The horizontal circle is numbered in a clockwise direction when looking down on the instrument. As the verniers occur on the outside edge of the upper plate the numbers increase from right to left so the scale is read in that direction.

Notes. (a) Magnifying lenses fixed over the vernier can be adjusted for focus.
(b) The readings must be made while looking straight down at the scales.
(c) Taking the mean reading of the opposite vernier scales provides greater accuracy.

Disadvantages of a vernier theodolite
The vernier theodolite was found to be large and heavy and the vernier scales difficult to read in poor weather conditions.

Optical theodolites
These instruments have glass horizontal and vertical circles which, compared with those of the vernier theodolite, are smaller, more accurate and easier to read. They are totally enclosed so that they are protected from dust and the effect of weather. The readings are made optically through a microscope parallel to the telescope. Internal prisms enable parts of both the horizontal

and vertical circles to be seen at the same time. Daylight is reflected to the circles for illumination or a torch can be shone through the lighting window in poor light conditions.

Reading systems are of four types.

Index line reading

This is the simplest system and relies on magnification to enable the circles to be read against a central index line. Figure 5.5 shows the readings on a Vickers VII instrument. The vertical circle is graduated in 5 minute divisions. Readings are made directly using the index line and estimated between the graduations.

The index line reading system is used on instruments for low accuracy work.

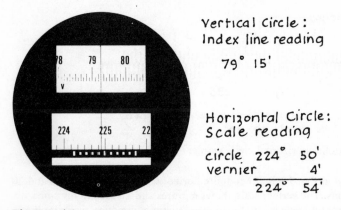

Vertical Circle:
Index line reading

79° 15'

Horizontal Circle:
Scale reading

circle 224° 50'
vernier   4'
          ‾‾‾‾‾‾‾‾‾‾
          224° 54'

Fig. 5.5 Vickers VII theodolite readings

Scale reading

In this reading system the circle graduations are read against a fixed scale. The scale may be a linear one or a vernier pattern. The horizontal circle in Fig. 5.5 shows a vernier scale enabling readings of 1 minute to be made. This is adequate for most building situations. Scales permitting readings to smaller values are fitted to some instruments. Figure 5.6 shows the Vickers VII theodolite.

Micrometer reading

Finer readings can be made with this system incorporating a micrometer control which, when turned, deflects the circle image and the micrometer scale image.

With the micrometer scale set to zero consider a reading showing the fixed index line falling between two circle graduation marks. The micrometer control is turned to displace the circle until a graduation mark coincides with the index line. The displacement made is recorded on the micrometer scale which is added to the circle scale showing at the time. The addition of these two readings is the total reading for the angle.

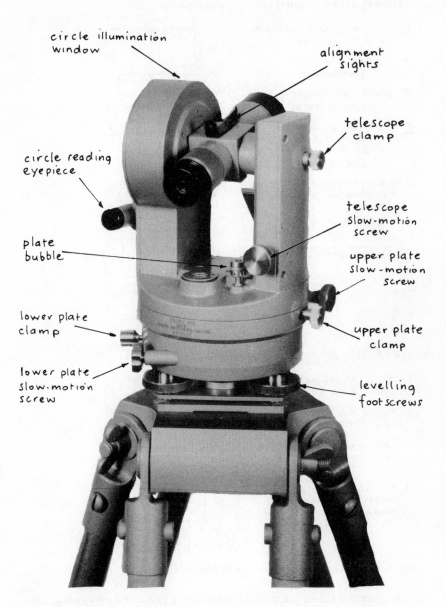

circle illumination window

alignment sights

telescope clamp

circle reading eyepiece

telescope slow-motion screw

plate bubble

upper plate slow-motion screw

lower plate clamp

upper plate clamp

lower plate slow-motion screw

levelling footscrews

Fig. 5.6 Vickers VII theodolite

88

Horizontal angle readings

1. Zero micrometer scale. Zero H-H scale.

2. Sight on target A. (reading a)

3. Sight on target B. (reading b)

4. Turn micrometer knob until H-H scale moves to 20 minute mark. (reading c)

5. Add micrometer scale reading to H-H scale reading.

$$
\begin{array}{rrr}
51° & 20' & \\
& 12' & 40" \\
\hline
51° & 32' & 40"
\end{array}
$$

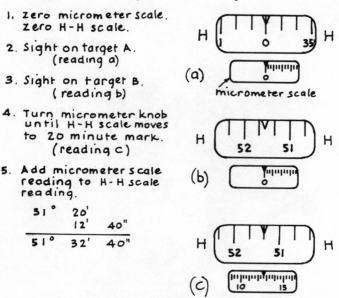

Setting-out horizontal angle of 128° 47' 20"

1. Zero micrometer scale. Zero H-H scale.

2. Sight on target A (reading a)

3. Set micrometer scale to 7' 20" (reading d)

4. Swing upper plate until H-H scale shows 128° 40' (reading e)

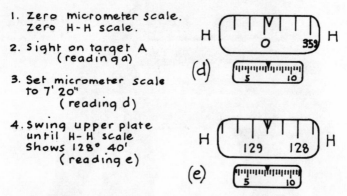

Fig. 5.7 Watts microptic theodolite ST456 readings

Figure 5.7 shows the method of reading a horizontal angle using the Watts microptic theodolite ST456 with readings made direct to 20 seconds. These instruments are suitable for medium accuracy work. Figure 5.8 shows the Watts microptic theodolite ST456. Digital reading micrometer theodolites show figures for every reading on the micrometer scale.

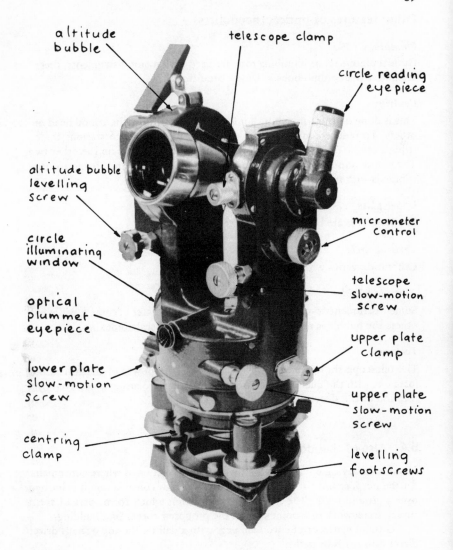

altitude bubble

telescope clamp

circle reading eyepiece

altitude bubble levelling screw

circle illuminating window

optical plummet eyepiece

lower plate slow-motion screw

centring clamp

micrometer control

telescope slow-motion screw

upper plate clamp

upper plate slow-motion screw

levelling footscrews

Fig. 5.8 Watts microptic theodolite ST456

Coincidence micrometer readings

To compensate for errors due to eccentricity of the circle in high accuracy work the mean reading at opposite sides of the circle is taken. A reading system is used which shows the two opposite graduations in the microscope at the same time. Turning the micrometer control enables the mean reading to be taken.

Direct readings can be made to one second or smaller.

Other features of optical theodolites

Plumbing

Optical plummets or plumbing rods are used with many instruments, these are better than plumb-bobs in windy conditions.

Centring

This is done on the tribrach on some instruments and at the tripod head on others. To reduce centring errors when measuring between stations three tripods can be used with the theodolite on one and targets on the other two. Theodolite and target sets are made for removal from, and fitting to, the tribrachs without disturbing the centred position.

Upper plate

The standards are hollow and carry the optical prisms.

Vertical circle

Calibration varies with instrument.

Altitude bubble

Some instruments carry coincidence bubbles for greater accuracy, while on others the bubble is replaced by automatic vertical scale indexing.

Telescope

The telescope shows inverted or erect image. The power of magnification increases with the quality of the instrument. Image focusing is achieved by rotation of the sleeve on the telescope body.

5.4. Use of theodolite

In some surveying operations the theodolite is positioned where convenient for the purpose but for most work connected with construction it is located over a ground mark. The mark can be a station which forms part of the survey framework or a point such as the proposed corner of a building.

Ground marks can be wooden pegs with a nail on the top or nails driven direct into roads or paths.

Setting-up the theodolite over a fixed point

Figure 5.9 shows each stage using a theodolite with an optical plummet.

Positioning the tripod

1. Loosen the tripod head wingnuts, extend the telescopic legs and set up the tripod over the ground mark with the head horizontal. Some tripods have head circular bubbles.
2. Suspend a plumb-bob from the centre of the head. There is usually provision for this but if not loop the cord over a pencil laid horizontally.

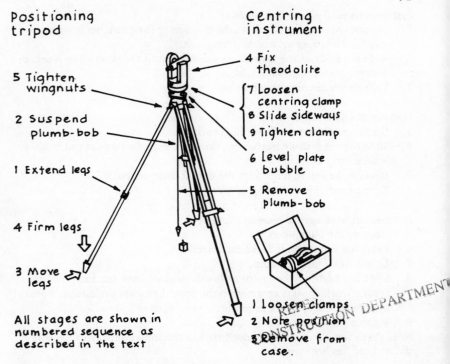

Positioning tripod

5 Tighten wingnuts

2 Suspend plumb-bob

1 Extend legs

4 Firm legs

3 Move legs

All stages are shown in numbered sequence as described in the text

Centring instrument

4 Fix theodolite

7 Loosen centring clamp
8 Slide sideways
9 Tighten clamp

6 Level plate bubble

5 Remove plumb-bob

1 Loosen clamps
2 Note position
3 Remove from case.

Fig. 5.9 Setting-up optical theodolite fitted with optical plummet

3. The plumb-bob should be horizontally within 10 mm of the ground mark, if not, alter the position of the tripod by moving each leg separately the same amount on the ground.
   Where the tripod has a centring rod it should be positioned on the ground mark and the tripod adjusted until the circular bubble is roughly centred.
4. Press the legs firmly into ground.
5. Tighten the head wingnuts, remove the head cap and place in the theodolite box or on the tripod leg fastening designed for the purpose.

Centring the instrument
1. While the theodolite is still in its case verify that horizontal circle clamps are loose.
2. Note which way it should be replaced in the case.
3. Remove the instrument from the case, holding it by the standard and tribrach.
4. Holding the standard with one hand screw to the tripod head with the other hand.

Instruments with plumb-bob centring

5. Loosen the centring clamp on the tribrach or for instruments with tripod head centring loosen the fixing bolt.
6. Slide the body of the instrument sideways until the plumb-bob is immediately above the ground mark.
7. Tighten the centring clamp or bolt.

Instruments with rod centring

5. Unclamp or unscrew the sliding provision.
6. Slide the head of the instrument sideways until the circular rod bubble is accurately centred.
7. Revolve the rod to make sure the bubble stays central.
8. Tighten the centring clamp or bolt.

Instruments with optical plummet centring

5. Remove the plumb-bob.
6. Level the plate bubble (described later).
7. Loosen the centring clamp.
8. Slide the body of the instrument sideways without rotation until the ground mark appears centrally in the optical plummet telescope. Focus by moving the eyepiece in or out.
9. Tighten the centring clamp.

Note that a few theodolites are not fitted to the tripod until after the tripod sliding head has been centred.

Levelling the instrument

The plate bubble is levelled by the three footscrews as described for the dumpy level and shown in Fig. 4.5.

If, after centring the bubble in two directions at 90° to each other, it is found that the bubble moves off centre when moved in any other direction, then permanent adjustment to the bubble setting is required. However, until this can be carried out the instrument can be satisfactorily used by removing half the error with the footscrews. The bubble should then stay in the same off-centre position whichever way it is directed.

Eliminating parallax

The cause of parallax has been explained in Chapter 4 in connection with levelling. The fault must be removed from all sighting instruments having diaphragm crosshairs if accurate readings are to be made.

1. Remove the cap from the object lens and direct the telescope to the sky.
2. Turn the eyepiece until the graticule lines appear sharply in focus.
3. Direct the telescope to a distant vertical object and turn the image focus control knob or sleeve until the object appears distinct.
4. When the eye is moved from side to side the graticule lines must not appear to float in relation to the object. If they do parallax is present and must be removed by repeating stages 1 to 4.

Measuring horizontal angles

It is normal practice for angles to be measured in a clockwise direction.

Consider three points on the ground A, B and C with the theodolite set up at C. In order to read angle ACB it is necessary to carry out the following sequence of operations.

Prepare the instrument

1. Where the instrument has an illuminating mirror, adjust it until maximum light appears in the circle reading microscope.
2. Turn the microscope eyepiece or the vernier magnifiers until the image is sharp.
3. Zero the micrometer scale.

Zero the circle

1. Rotate the upper plate over the circle until the index mark shows nearly 0°.
2. Tighten the upper clamp.
3. Turn the upper slow-motion screw to obtain an exact reading of 0°.

Sight on target A

1. Using the alignment sights direct the telescope towards target A.
2. Tighten the lower clamp.
3. Focus the telescope on target A.
4. Turn the lower slow-motion screw so that the graticule vertical line bisects target A.

Sight on target B

1. Loosen the upper clamp.
2. Swing the upper plate in a clockwise direction until the telescope alignment sights point at target B.
3. Tighten the upper clamp.
4. Focus on target B.
5. Turn the upper slow-motion screw so that the graticule vertical line bisects target B.

Read the angle

To read the angle follow the procedure set out in the section 'Types of instrument' earlier in this chapter.

Setting-out horizontal angles

Carry out the procedure described in 'Measuring horizontal angles' up to and including 'Sight on target A' and continue as follows:

1. Instruments with micrometers, set as shown in Fig. 5.7(d).
2. Loosen the upper clamp.
3. Swing the upper plate in a clockwise direction reading the horizontal circle at the same time.
4. When nearing the required reading tighten the upper clamp.

5. Turn the upper slow-motion screw until the reading is shown by the index line.
6. Tilt and focus the telescope to where the target is to be positioned.
7. An assistant moves target B sideways until it coincides with the vertical graticule line.

Methods of angle measurement

Where several angles are measured from the same point the horizontal circle remains clamped in position and the angles measured in sequence in a clockwise direction from a reference object (RO). This is called a round of angles. Angles should always be measured more than once so that gross errors are detected.

The first reading need not be made a value such as $0°$ or $180°$. The reading which happens to occur when the instrument is set-up can be used and the second measured reading deducted from it.

Face left, face right

The theodolite telescope can transmit through $360°$ in the vertical plane. The normal position for first reading is for the vertical circle to be on the left as the surveyor looks through the telescope, this is called face left. When the surveyor transits the telescope $180°$ and again looks through the telescope the vertical circle is on his right. This position is face right. Figures 5.1 and 5.9 show the instrument in face left position. Conventionally, the rule used by surveyors is face left — swing right, face right — swing left. This allows for play in the bearings of the instrument but when using modern theodolites for normal accuracy work it is satisfactory to swing right (clockwise) for each reading.

Face left and face right readings should be taken for each angle measured, as the mean of the readings compensates for certain instrumental errors. It will also show up gross errors.

Simple reversal

This method of measurement is recommended for all work requiring a moderate degree of accuracy. It combines compensation for certain errors by using both faces of the instrument but with readings taken on opposite sides of the circle which compensates for circle eccentricity, see Fig. 5.10(a).
1. Set the horizontal circle index to zero.
2. Measure and book the angle (face left).
3. Transit the telescope.
4. Set the horizontal circle index to $180°$.
5. Measure and book the angle (face right).
6. Take mean reading.

Reiteration

In high accuracy work a number of readings on both faces are taken on different parts of the circle, for example one on each quadrant, and the mean reading found. The greater the accuracy required the more readings are taken.

This method is used with precision instruments on large triangulation schemes. See Fig. 5.10(b).

Repetition

To increase the accuracy of a single reading a multiple of the measured angle is read and the mean found. This is carried out in the following order:

1. Measure the horizontal angle ACB but do not take the reading.
2. Loosen the lower clamp and again sight on target A.
3. Measure the angle ACB a second time. The angle now recorded will be approximately double the angle ACB.

This can be done a number of times and the total angle divided by the number of times the measurement was repeated. Readings are taken on both faces. This method is used particularly where high accuracy is required for measuring very small angles. See Fig. 5.10(c).

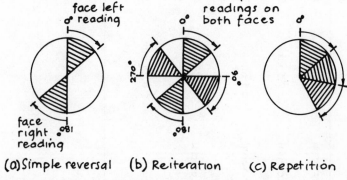

(a) Simple reversal (b) Reiteration (c) Repetition

Fig. 5.10 Methods of horizontal angle measurement

Accuracy

The degree of accuracy required of angular measurements depends on the type of survey being undertaken. The degree of accuracy achieved depends on the quality of instrument used and the method employed. Table 5.1 has been prepared to give an indication of the accuracy obtainable with different types of theodolite. It has been compiled partly from the permissible deviations shown in BS 5606:1978 *Code of practice for Accuracy in Building*.

Booking readings

A method of booking angles using the simple reversal method is shown in Table 5.2. Entries are made in pencil and mistakes erased. Zeroes are shown in front of single figures in the minutes and seconds column. Calculations are completed before the instrument is moved. With an instrument in good condition the difference between face left and face right readings should not exceed 1 minute and should normally be less than this amount. If a greater difference is recorded the angle should be remeasured.

Table 5.1 Accuracy of typical theodolites

Type	Calibration	Method and accuracy	Deviation at 50 m	Sight length at which 10 mm deviation occurs	Application
Vernier	20 second	Simple reversal ± 40 secs	± 10 mm	52 m	Low accuracy building
Optical scale reading	1 minute estimated to 30″	Simple reversal ± 30 secs	± 7.5 mm	69 m	Low accuracy building
Optical micrometer reading	20 seconds estimated to 5″	Simple reversal ± 20 secs	± 5 mm	103 m	Medium accuracy building and engineering
Coincidence micrometer reading	1 second estimated to 0.5″	Simple reversal ± 5 secs	± 1.25 mm	413 m	Medium accuracy engineering and land surveys
Coincidence micrometer reading	1 second estimated to 0.5″	Reiteration ± 1 sec	± 0.25 mm	2063 m	High accuracy land surveys

Table 5.2 Simple reversal booking

Instrument	Target	Face left		Face right		Mean of difference	Remarks
		Reading	Difference	Reading	Difference		
A	B	0° 00' 00"		180° 00' 00"			
	C	63° 24' 20"	63° 24' 20"	243° 24' 40"	63° 24' 40"	63° 24' 30"	
	D	114° 07' 00"	114° 07' 00"	294° 07' 40"	114° 07' 40"	114° 07' 20"	

(First readings need not start at exactly 0° and 180°)

5.5. Fieldwork notes

Position of instrument

The theodolite is invariably positioned over an existing ground mark. Where there is a choice of positions the following situations should be avoided.
1. Soft, marshy and unstable ground.
2. Public or site roads where the instrument is exposed to damage.
3. Close proximity to plant or vehicles causing vibration.
4. Where observation is likely to be interrupted.

Use of instrument

1. When measuring angles do not touch the tripod.
2. Do not use force on any of the controls.
3. Slow-motion screws often have limited travel. Continuous turning will use up all the thread and cause jamming.
4. Do not leave the instrument unattended.
5. To move the instrument more than a few metres, the theodolite should be unscrewed from the tripod and carried in its case.

Weather

The remarks made on p. 75 concerning the level also apply to the theodolite.

Targets

These must be vertical and positioned exactly centrally over the ground marks. The telescope should focus low down on the target, direct to the ground mark if possible. The width of the target depends on the distance from the theodolite.

Ranging poles

Are satisfactory for sights of more than 25 m and should be supported in a tripod stand.

Arrows

Can be used for close sights and held vertically over the ground marks while observations are made.

Tripods

Can be positioned over the ground marks and sightings made to the plumb-bob or string. The tripod legs will have to be positioned not to obstruct the sight lines.

Three tripod system

This is where targets and theodolites can be moved from one tripod to another without disturbing the tribrach centring. Targets are painted red and white to form a central cross.

Figure 5.11 illustrates the targets described.

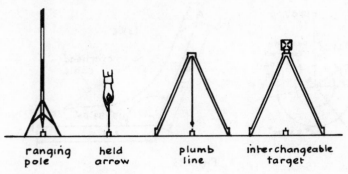

Fig. 5.11 Types of target

5.6. Errors with theodolite

Gross errors

(a) Inaccurate centring of instrument over ground mark.
(b) Incorrect levelling of instrument.
(c) Parallax.
(d) Inaccurate sighting on targets.
(e) Misreading angle.
(f) Booking errors.
(g) Calculation errors.

Remedy

Take care when setting-up the instrument and compare several readings.

Systematic errors

These are errors caused by instruments being out of permanent adjustment.

Remedy

Permanent adjustments are not dealt with in this book. Most instrumental errors are eliminated by reading both faces and different parts of the circle as in simple reversal, or for high accuracy work the reiteration method.

5.7. Questions

Formulae shown in Appendix 1 (p. 154).

Qu 5.1 Figure 5.12 shows details of angles to be set-out from point A. With the instrument set to zero on the reference object B calculate the angles of lines AC, AD, AE, AF as bearings measured around the whole circle.

Qu 5.2 Points A and B in Fig. 5.13 mark where towers are to be erected to support an overhead power cable. Calculate length AB. Use the cosine rule.

Qu 5.3 The width between fences on a road cutting shown in Fig. 5.14 is to be found. Points A and B are positioned parallel to the south fence. Use the sine rule first.

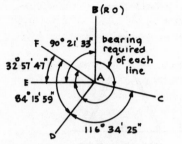

Fig. 5.12

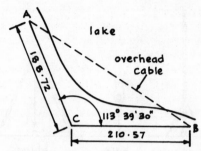

Fig. 5.13

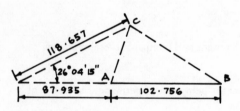

length AB = 38·56
angle CAB 74° 09'
angle ABC 59° 37'

Fig. 5.14

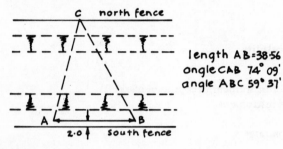

Fig. 5.15

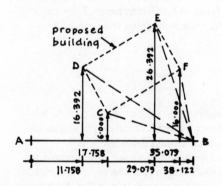

Fig. 5.16

Qu 5.4 Point C in Fig. 5.15 is to be set-out by angles using two theodolites, one at station A and one at station B. An assistant will move the target until it aligns with both theodolite sights. Calculate angles CAB and ABC. Use the cosine rule .

Qu 5.5 Figure 5.16 shows sizes for setting-out a proposed building using running lengths on line AB and square offsets from AB. It has been decided to set-out with polar co-ordinates from B using A as the reference object. Calculate all necessary bearings and lengths.

Answers shown in Appendix 2 (p. 156).

Chapter 6

Surveys of buildings and small sites

Building surveys take the form of either:
(a) a written report describing the nature and present condition of the fabric; or
(b) a scale drawing prepared from site measurements and notes.
This chapter deals only with part (b).

6.1. Purpose

Measured surveys are required for alterations, extensions, valuation or record purposes.

The specific reason for the survey must first be established as this will influence the type of information to be obtained on site. For example, if the purpose of the survey is to determine what settlement has taken place the surveyor will need to record and show on the drawing the position and magnitude of cracks that are evident. The surveyor should be told which of the following services are to be shown:
(a) hot and cold water supply;
(b) waste plumbing and drainage;
(c) electric, gas and oil installations.
Sanitary fixtures should always be shown.

Where the drawings are required for interior design purposes all unfixed items of equipment such as cookers and refrigerators should be shown together with furniture.

6.2. Instruments

Measuring and marking

Tapes

A 20 m or 30 m coated glass-fibre tape should be used. Steel tapes can easily be damaged and are not sufficiently flexible to measure into the corners of a room. The only occasion when steel tapes are used is for structural engineering work where very accurate measurements are necessary to determine the required sizes of steel beams or precast concrete units.

A flexible steel pocket tape 2 or 3 m in length is very useful for taking small measurements.

Folding rods

These are made from hardwood in single-fold or multi-fold patterns which, when opened up, form a rigid rod 2 m long. To conform to BS 4484:1969 the upper edge is graduated in 1 mm divisions and the lower edge in 5 mm divisions. See Fig. 6.1.

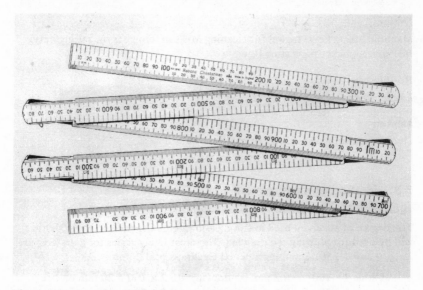

Fig. 6.1 Hardwood multi-fold measuring rod

Ranging poles

Used to position control points for measurements of site. Stands are used for hard surfaces.

Arrows

Used for marking end of tape for site measuring. Arrows and ranging poles are described in Chapter 2.

Sketching

Paper

Use a hardback book, or use cartridge or tracing paper fixed to a clipboard. Faint printed squared paper can be used as a guide for entries or tracing paper can be fixed over the top. A large clear plastic cover sheet is useful for protecting the paper when working in the rain.

Pencils

Use HB grade and use coloured pencils for showing services. Carry an eraser and a sharpener.

Straight edge

While no attempt is made to draw sketches on site to scale a 150 mm scale rule is useful for drawing straight lines.

Other equipment

Various items are necessary for the following situations.

Drains

Manhole lifting keys; trowel for cleaning MH rim; crowbar for raising heavy MH cover; bucket for testing flow.

Roof space

Ladder; torch.

Marking

During measuring use crayon or chalk.

6.3. Method of measurement

Degree of accuracy

The degree of accuracy used in site measurements depends on the scale that will be used for plotting the drawing. The most likely scales for a drawing are 1 : 100 and 1 : 50. Assuming a pencil thickness of 0.2 mm:

at scale 1 : 100   0.2 mm on drawing = 0.2 x 100 = 20 mm on site
at scale 1 : 50   0.2 mm on drawing = 0.2 x 50 = 10 mm on site

For normal work site measurements should be taken to the nearest 10 mm and rounded up or down accordingly. The only exception is where precast concrete, structural steelwork, metal components, machinery or joinery is to be fitted between faces in the building. Here the measurements should be taken to the nearest mm and the sizes shown on the drawing.

Separate measurements (Fig. 6.2(a))

The two disadvantages of separate measurements are that:
(a) any mistakes will cause other measurements to apply at the wrong positions;
(b) if, by coincidence, most measurements in a row are rounded up or rounded down the effect would be a cumulative error that would only show up if an overall measurement were taken. An example of this is in Fig. 6.2(a) where the two runs of separate measurements when totalled do not agree with the two overall lengths shown in Figs. 6.2(b) and (c).

Running measurements (Figs. 6.2(b) and (c))

The two advantages of running measurements are that:
(a) one mistake will not effect any other measurement;
(b) if measuring to the nearest 10 mm, every measurement is within 10 mm of its true length.

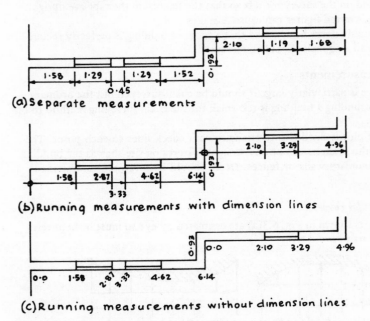

Fig. 6.2 Methods of recording measurements

Method used

Running measurements should be used wherever possible. They should be as long as the feature allows. Where, for example, a wall projects forward, or is stepped back as in Fig. 6.2(b), another start to running measurements will be made.

There are occasions, particularly with vertical work, when running measurements cannot be used. In this case some method of taking an overall measurement must be devised to provide a check on accuracy.

Where dimension lines are not used in running measurements care must be taken to show the figures against the feature being measured as in Fig. 6.2(c). Dimension lines need not always be used with separate measurements in which case the figures are shown centrally between the points being measured. If it is not perfectly clear where the figures apply then a dimension line should always be used to avoid ambiguity.

6.4. Taking measurements

Wall measurements

Running measurements should be taken at approximately 1.25 m above the ground or floor to pass window openings. The tape is pulled tight and held horizontally. The surveyor takes the readings so his assistant holds the metal tape loop against the corner of the wall. The beginning of the tape is positioned on the surveyor's left so that the figures on the tape are upright. Reading inverted figures can cause mistakes.

It should never be assumed that a room or building is perfectly rectangular. Both diagonals should be measured.

Site measurements

If the site is particularly large it should be chain surveyed but the ordinary plot surrounding a building is too small for the chain surveying method to be used.

The plot is divided into triangles with check lines to each point. The sides of the triangles can be any straight features on site such as the building walls, boundary walls or fences. Detail should be measured as described below.

Non-circular curves

The offsets shown in Fig. 6.3(a) are estimated by eye so must be kept relatively short to avoid error.

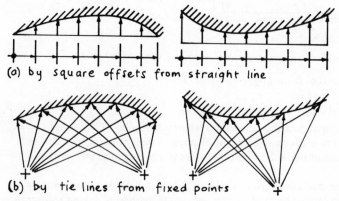

(a) by square offsets from straight line

(b) by tie lines from fixed points

Fig. 6.3 Measuring non-circular curves

The tie lines is Fig. 6.3(b) are taken from either two existing features or two points fixed for the purpose, these points being fixed by tie lines to existing features. The distances between points on the curve are at estimated regular intervals and marked by chalk or crayon.

Circular curves

If the curve is known to be circular the radius can be found graphically as in Fig. 6.4(a) or by calculation as in Fig. 6.4(b). Where the graphical method is adopted the lengths AB, BC and AC are measured on site and the triangle drawn to scale with at least two of the sides bisected to find the centre of the circle.

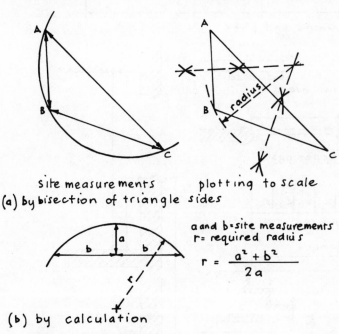

site measurements   plotting to scale

(a) by bisection of triangle sides

a and b=site measurements
r= required radius

$$r = \frac{a^2 + b^2}{2a}$$

(b) by calculation

Fig. 6.4 Measuring circular curves

In-line measurements (Fig. 6.5(a))

These are useful for fixing the position of detail and, combined with other measurements, can also fix the direction of lines.

Recording angles

A protractor is too small to measure angles on site. The method shown in Fig. 6.5(b) is normally used with the lengths being as large as is convenient.

Columns and pipes (Fig. 6.5(c))

Measurements are taken to centre-lines and the diameter of round objects found by dividing the measured circumference by π.

108

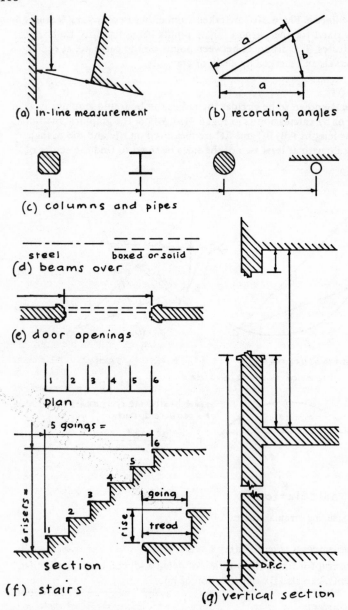

(a) in-line measurement

(b) recording angles

(c) columns and pipes

steel   boxed or solid
(d) beams over

(e) door openings

plan

5 goings =

section

going
rise   tread

(f) stairs

(g) vertical section

D.P.C.

Fig. 6.5 Measuring detail

Overhead features

Room plans are always shown sectionally. Features above the section plane such as trap-doors and concrete or boxed steel beams are shown with a broken line. Open steel beams or trusses are shown with a chain line. In each case a description is shown and followed by the word 'over'. See Fig. 6.5(d).

Door openings

The measurements to doors in external walls are taken to the brick opening. The width of the door is also measured. The measurements to internal doors are taken to the door linings as shown in Fig. 6.5(e). The size of the architrave is measured separately.

Flight of stairs

It is inaccurate to measure one step and multiply this by the number in the flight because of cumulative errors. The total rise and going for the whole flight should be measured and the size of each step found by division. To obtain the measurement for the total going it may be necessary to suspend a plumb-bob to the nosing of the bottom step and measure to the string. The risers should be numbered on section and plan as shown in Fig. 6.5(f).

Vertical measurements

Room heights are shown on plan with a circle drawn around the figures.

As the ground level will vary around a building external heights are taken from a horizontal datum, normally the damp-proof course or a projecting plinth.

The ground surface is measured down from the datum at several points around the building. Each floor must be related to the datum by an external measurement as shown in Fig. 6.5(g). Check measurements are possible internally by finding the floor thicknesses at stair wells.

Wall thickness

The thickness of walls is found through open doors or windows. The type of brick bonding used is a good indication of whether the wall is solid or of cavity construction.

Roof

Entry to the roof space will allow the internal height to be measured so the pitch can be found. The roof construction should be noted and the position and size of cold water cisterns.

Externally the projection of chimneys above the roof line can often be found by counting the courses and multiplying by the single brick height including the bed joint.

Recording measurements

By the time the survey is complete there may be hundreds of measurements and notes recorded and great care must be taken to maintain the clarity of each entry.

Wherever possible measurements should be shown without dimension lines as too many lines on the sketches cause confusion. Where dimension lines are used the arrowheads must extend to the full length of the measurement.

On the sketch always enter the readings close to the detail being measured and in the direction that the surveyor stands when viewing the tape. This means that figures will appear in all directions on the sketch. Figures, dimension lines and notes must not obscure other information. If the entry is not perfectly clear or confuses other entries it should be erased and re-written.

All measurements necessary to complete the scale drawing must be anticipated when on site. Check measurements should be taken at all times.

6.5. Notes and sketches

Construction notes

At the same time as measurements are being taken the type of construction should also be noted. Abbreviations are normally used to save time and space. Such abbreviations can be those recommended in BS 1192 or ones devised by the surveyor.

Sketches

Sketches should be clear and to proportion. They can be free-hand or a pocket straight edge can be used. Sketches should be as large as possible to fit on the sheet. Extensive work can extend over more than one sheet but certain points must be reproduced on adjacent sheets to provide continuity of measurement. Complicated parts should be sketched separately at a larger size.

6.6. Order of work

Preliminary investigation

Before visiting the site make reference to the up-to-date OS 1 : 1250 sheet. If earlier sheets are also available they could indicate when extensions took place which could account for variation in constructional methods.

Make enquiries to find if any drawings of the building already exist. Old out-of-date drawings can be used as a basis for the present building and save a considerable amount of time on sketching and measuring on site.

The site

Walk over the site to find the shape and then sketch the boundaries, fences, gates, paths and buildings. Show dimension lines to divide the site into triangles with check lines to each point. Take measurements to complete each part before moving on to the next part. Make notes on construction of fences and paths, at the same time measure manholes for size, location and direction. Remove the covers and show the arrangement of channels. Show gullies, rainwater downpipes and soil and vent pipes and indicate pipe run with chain lines. Note the diameter of the drain, the direction of flow and whether for soil or surface water.

If the drain connections are not evident tip water down one fitment at a time to find where it discharges. Coloured dyes help in tracing drain runs. Where the north point cannot be established from an OS sheet take a magnetic bearing on site using a prismatic compass. Site plan measurements are shown in Fig. 6.6 but this example does not include drainage.

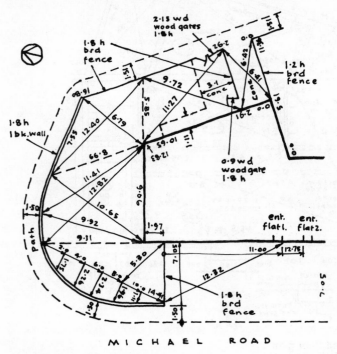

Fig. 6.6 Site plan measurements

The building

The recommended order of work is described in sequence. In all cases sketch the view first and then add measurements and notes. Photographs can be taken to augment sketches. Figures 6.6, 6.7, 6.8, 6.9, and 6.12 all relate to the same site and building.

1. Ground floor plan

Sketch and measure the outside of each external wall, showing windows, doors, steps and pipes. Construction notes should be added at the same time. While outside decide where the cross-section should apply to show the roof construction.

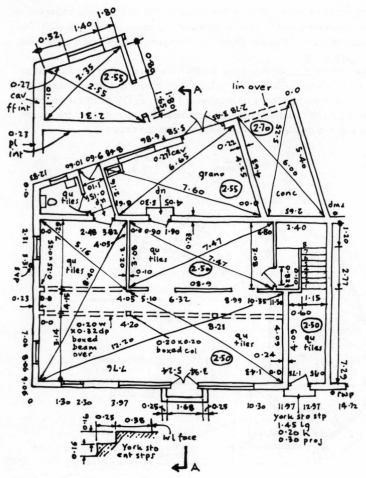

Fig. 6.7 Ground floor plan

Inside the building show walls, fireplaces, chimney breasts, door swings, stairs, beams over and such fixtures as sinks, washbasins, baths and WCs. The same symbols can be used as shown in Fig. 6.10. The position of the vertical section must be shown with arrowed lines on plan.

An example of ground floor plan measurements is shown in Fig. 6.7.

2. Ground floor vertical section

The purpose of a section is to show constructional detail. Vertical sections are generally taken at right-angles to the roof ridge so that the roof construction can be detailed. To show construction the sections should pass through door and window openings. Sections do not have to be taken in one flat plane but can be stepped to cut through the detail required.

113

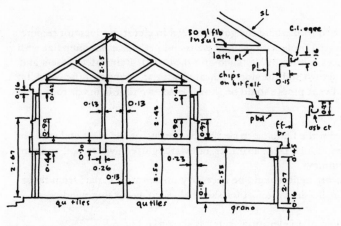

Fig. 6.8 Section A — A

The ground floor vertical section should show the floor level in relation to the outside ground level. To show construction of a suspended ground floor a board would have to be raised to reveal the sleeper walls etc.

All wall thicknesses should be measured and construction notes shown.

An example of vertical section measurements is shown in Fig. 6.8.

3. Basement
If the building has a basement it should be measured at this stage to show plan and vertical section.

4. First floor plan
If tracing paper is being used for sketches, the outline of the external walls can be traced from the ground floor plan. The same type of information is required as for the ground floor. A note should be made of stud partitions which can be detected by the hollow sound made by tapping the surface. The directions of the floor joists should be indicated and trap-doors in ceilings measured and noted 'over'.

5. First floor vertical section
This is a continuation of the ground floor section so must be taken on the same cutting plane as in Fig. 6.8.

6. Roof space plan
Access is gained by ladder through a trap-door. A torch will be necessary to see the position of chimney and cold water cisterns.

7. Roof space vertical section
Continuing the vertical section already measured, show the construction of the roof together with the horizontal spacing of the trusses and rafters. All timber section sizes must be measured and any boarding or underfelt noted. Type and thickness of insulation should be recorded. Refer to Fig. 6.8.

8. Roof plan

A simple pitched roof where the gutter is seen in elevation does not require a plan. Complicated roofs with parapet walls and perhaps combining flat with pitched surfaces should be sketched and measured. Rainwater gutters and outlets to be shown and the slope indicated with an arrow and the word 'fall'. Chimneys and vent pipes should be shown, also any roof lights or roof access.

9. Elevations

Stand back from the building to sketch the outline of the walls and roof and show windows, doors, DPC, airbricks, rainwater gutters, waste, soil and vent pipes, and chimneys.

Few measurements should be necessary on elevations apart from those that could not be shown on other sketches, such as varying window sill heights.

An example of an elevation sketch is shown in Fig. 6.9.

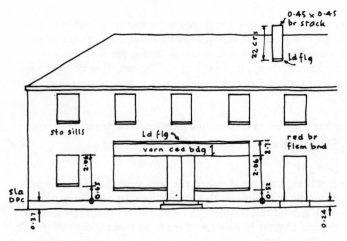

Fig. 6.9 Front elevation

6.7. Drawing to scale

The site

The site is plotted to a standard metric scale using compasses and verified by check measurements. A north point should be shown and a title panel.

The building

For the drawing to comply with BS 1192:1969 the views should be in first-angle orthographic projection. An explanation of this projection is illustrated in Fig. 6.10 which shows how a three-dimensional object is drawn in a two-dimensional plane.

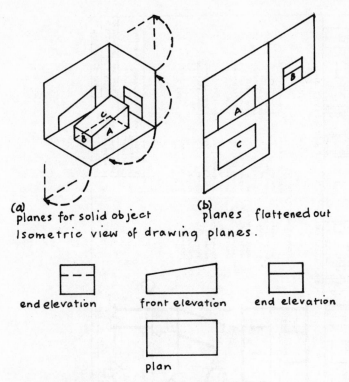

(a) planes for solid object   (b) planes flattened out

Isometric view of drawing planes.

end elevation   front elevation   end elevation

plan

Fig. 6.10 First-angle orthographic projection

Drainage		Heating	
Gully	⊡ G	Boiler	⊡ B
Intercepting trap	⊡ IT	Kitchen	
Manhole (soil)	⊡ MH	Sink	▣
Manhole surface water	○ MH	Sanitation	
Rainwater head	▭	Bath	▭
Rain water pipe	○ RWP	Bidet	∪ BT
Vent pipe	○ VP	Wash basin	▦
Water Supply		Shower unit	S
Cold water cistern	▢ CWC	Urinal	⌐
Hot water cylinder	○ HWC	WC	⊖
Indirect cylinder	◎		

Fig. 6.11 Selection of graphical symbols from BS 1192:1969

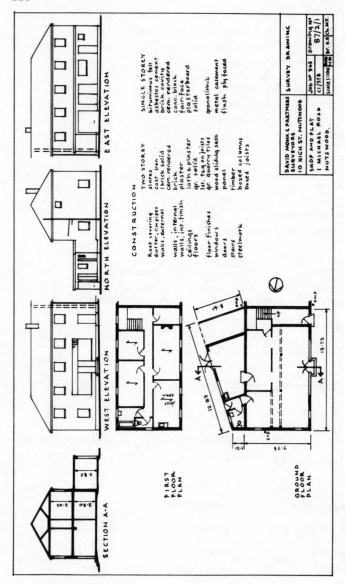

Fig. 6.12 Scale drawing

All elevations can be drawn using the same first-angle convention and the position of the drawn section is decided by the direction of the arrows shown on plan. A plan must be drawn at each floor and possibly the roof.

Graphical symbols from BS 1192:1969 shown in Fig. 6.11 should be used.

The overall dimensions should be shown. Running dimensions are not shown on scale drawings. Notes on construction should be collected in groups and not spread over the whole drawing.

Abbreviations shown in BS 1192 can be used. Some have already been given in Fig. 2.14. Other selected abbreviations are shown below.

Airbrick	AB	Finished floor level	FFL
Asbestos cement	abs ct	Granolithic	grano
Boarding	bdg	Ground level	GL
Cast iron	CI	Hardwood	hwd
Centre to centre	c/c	Insulation	insul
Centre-line	₵	Internal	int
Column	col	Joist	jst
Cupboard	cpd	Plasterboard	pbd
Damp-proof course	DPC	Polyvinyl chloride	PVC
Damp-proof membrane	DPM	Reinforced concrete	RC
Diameter	dia	Sewers foul	FS
Diameter, inside	I/D	Sewers surface water	SWS
Diameter, outside	O/D	Softwood	swd
External	ext	Tongue and groove	T&G

If the drawing is being prepared for a house owner who is not familiar with building terminology it is better not to abbreviate.

A north point should be shown on plan and a title panel to show all relevant information including the scale.

The example in Fig. 6.12 is the completed scale drawing of the building previously shown with site measurements. Abbreviations have purposely been avoided.

6.8. Errors

Gross errors

On site

(a) Taking insufficient measurements.
(b) Misreading the tape.
(c) Misbooking the measurements.
(d) Confusing mm and m.
(e) Unclear figures.
(f) Figures in the wrong position.

118

Drawing office
(a) Misreading the figures.
(b) Misreading where the dimensions apply.
(c) Mis-scaling.
(d) Inaccuracy in plotting.

Remedy
(a) Sufficient time should be allowed on site for the work to be carried out thoroughly.
(b) All the measurements required should be anticipated.
(c) Sufficient check points to be taken.

Systematic errors
(a) The tape measures incorrect length.
(b) Tape read when it is sagging.

Remedy
(a) Check tape against a standard known length as described for chain surveying.
(b) Apply correct tension on the tape and do not use it unsupported over a distance of 10 m.

6.9. Questions

Qu 6.1 (a) Three points A B C are marked on a circular curved road kerb. AB = 6.70 m, BC = 4.17 m, AC = 10.48 m.
Draw the triangle to a suitable scale and graphically find the radius of the kerb.

(b) Figure 6.13 shows a stone circular curved arch. From the sizes given calculate the radius.

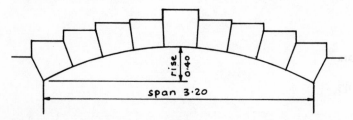

Fig. 6.13

Qu 6.2 The site of a house and garden is shown in Fig. 6.14. On this drawing show dimension lines which would enable the site to be plotted accurately to scale. There are various solutions, one of which is shown in the answers.
Qu 6.3 Figure 6.15 shows a pictorial view of a house. Sketch to proportion the front elevation, end elevation and roof plan in first-angle orthographic projection.

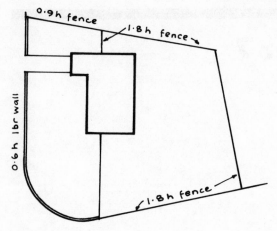

Fig. 6.14

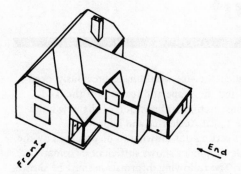

Fig. 6.15

Answers shown in Appendix 2 (p. 156).

Chapter 7

Setting-out

When the setting-out stage is reached the site would have previously been surveyed and probably levelled to find the shape and features of the land. Drawings showing the proposed construction and position of the new building will have been submitted to the Local Authority and approved.

A large development requires a setting-out drawing to be prepared by the architect or engineer, but a small development shows sufficient information on the block plan and general plan. The following information must be shown on drawings.

1. Position of proposed building(s) in relation to existing boundaries or roads.
2. Position and levels of new roads and drains.
3. Levels of new building at foundations, floors etc. Small buildings may only show the vertical dimensions between these items.

Setting-out is done by the contractor and may be checked by the architect or engineer.

Before any setting-out is started the site drawing should be 'proved'. This means that the shape of the site and the levels shown must be checked and proved correct. Existing horizontal control points should be re-located and existing reference levels verified.

To standardize setting-out methods the recommendations shown in the *CIRIA Manual of Setting-out Procedures* have been adopted where they apply to the scale of work covered in this book. It is recommended that surveyors involved in setting-out should obtain a copy of the manual which is published by the Construction Industry Research and Information Association. Copies are obtainable from Pitman Publishing Ltd, Pitman House, Parker Street, Kingsway, London WC2B 5PB.

7.1. Instruments for measuring and marking

Steel tape

To comply with BS 4484:1969 coated steel tapes should be 10, 20 or 30 m long. They should be graduated at 5 mm intervals throughout apart from the first and last metres which show 1 mm graduations. Steel tapes should be checked against a standard length from time to time. See 'Standardizing', p. 20, Chapter 2. A steel tape is shown in Fig. 7.1.

Fig. 7.1 Steel tape

String line

The line stretched between fixed points can be bricklayers' cotton, hemp or nylon line. The last has the advantage of not absorbing moisture and so will remain in correct tension at all times.

Timber

Pegs should be 50 mm x 50 mm x 500 mm softwood planed with a point cut at one end; stakes to be the same as pegs but 1500 mm long. Rails and boards should be 25 mm x 100 mm and up to 1500 mm long.

7.2. Instruments for forming right-angles

Builders' squares

Made from softwood, they may be of a temporary nature made quickly on site with lapped joints, or they may be intended for regular use and have halved joints. In both cases they are set-out using a 3, 4, 5 triangle. See Fig. 7.3.

Fig. 7.2 Site square

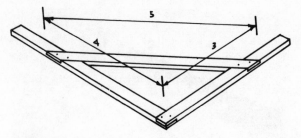

Fig. 7.3 Builders' square

Site square

This instrument has 2 fixed-focus tilting telescopes at right-angles to each other as shown in Fig. 7.2. A circular bubble is built into the top of the instrument. It is mounted on a tripod which has a plumb rod attached.

Method of use

1. Set up the instrument over the fixed peg A so that the plumb rod touches the nail in the top of the peg.
2. Adjust the sliding legs until the circular bubble is central.
3. Sight through one telescope towards the fixed peg B turning the instrument until the crosshairs bisect the nail. See Fig. 7.4(a).
4. Without rotating the instrument move to sight through the other telescope.
5. The assistant moves peg C sideways until it appears centrally in the telescope. See Fig. 7.4(b).
6. The assistant is directed to position the nail in the peg top so that it bisects the telescope crosshairs.

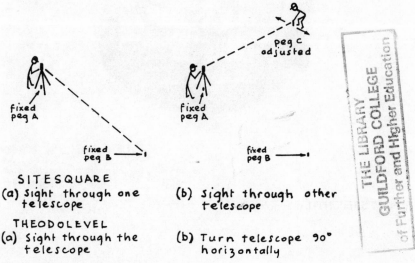

SITESQUARE
(a) Sight through one telescope

(b) Sight through other telescope

THEODOLEVEL
(a) Sight through the telescope

(b) Turn telescope 90° horizontally

Fig. 7.4 Use of site square and theodolevel

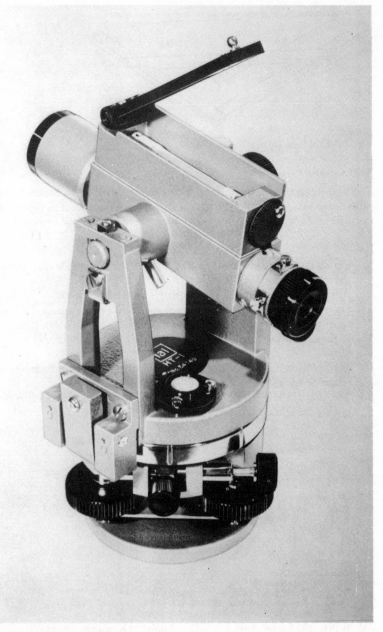

Fig. 7.5 Hall HT-1 theodolevel

Theodolevel

This multi-purpose instrument shown in Fig. 7.5 is useful for small building operations including setting-out. It carries a telescope with good magnification which shows an erect image.

The telescope can be clamped in the horizontal position and used as a tilting level. When unclamped it can be directed to sight on setting-out pegs and also used to plumb framed buildings from a distance. Another feature is its use for setting-out angles of 45° and 90°.

Method of use for setting-out right-angles

1. Set up the instrument over the fixed peg A so that the suspended plumb-bob touches the nail in the top of the peg.
2. Adjust the 3 screws to centralize the circular bubble.
3. Direct the telescope to focus on fixed peg B using the horizontal clamp and slow-motion screw until the crosshairs bisect the nail. See Fig. 7.4(a).
4. Turn the telescope horizontally until the 90° clickstop is located.
5. The assistant moves peg C sideways until it appears centrally in the telescope. See Fig. 7.4(b).
6. The assistant is directed to position the nail in the peg top so that it bisects the telescope crosshairs.

7.3. Instruments for levelling

Spirit level

An instrument made from hardwood or metal which incorporates one or more bubble tubes. The bubble tube is made from clear glass or plastic and is arched. It is partially filled with a coloured spirit and sealed leaving an air pocket which forms a bubble at the top. Spirit is used rather than water to prevent freezing. When the instrument is perfectly horizontal (for levelling) or vertical (for plumbing) the bubble will be centred between the etched marks. To check for accuracy reverse the spirit level on a flat surface; the bubble should stay central. A metal spirit level is shown in Fig. 7.6.

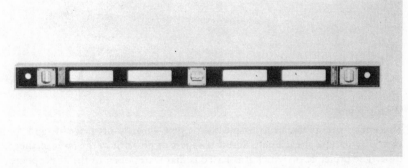

Fig. 7.6 Spirit level

Method of use

To transfer a level over a distance the spirit level is used in conjunction with a straight edge board. Working from the first peg which is at reference level the second peg is driven in until the bubble on the spirit level is central. To cancel errors the board and spirit level together are reversed each time when positioning the next peg as shown in Fig. 7.7. An accuracy of ±5 mm in a distance of 5 m is considered normal.

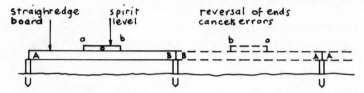

Fig. 7.7 Establishing a level line with spirit level

Line level

This is a short spirit level made in metal or plastic with hooks at each end as shown in Fig. 7.8. It is suspended centrally between supports on a tightly stretched string line.

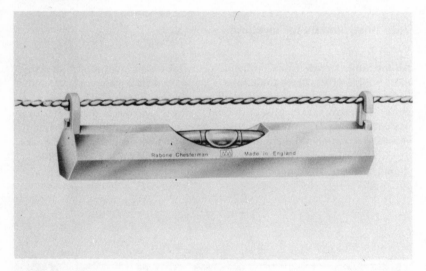

Fig. 7.8 Line level

Water level

This consists of a length of flexible plastic pipe with a short piece of rigid clear plastic tube at each end. Screw stoppers or plugs fit in the ends of the tube. The principle on which it is based is that water open to the atmosphere settles to the same level.

Method of use

1. Fill with water to half-way up the tubes avoiding air-locks.
2. Hold the tubes together. Water should stand at the same level. If it does not air-locks are present; these must be removed.
3. Fit both caps and carry to where required.
4. Tie one tube loosely at the reference level and hold the other tube next to it.
5. Remove both caps. Water will settle to the same level.
6. Check that the water level is at reference level and tie the tube securely.
7. With both caps removed move the unsecured tube to where the level is to be transferred, keeping the tube at the same level to prevent loss of water.
8. Mark off the required level which is the water level. Ignore tube graduations.
9. When all levels are marked off replace the caps.

Note: If part of the pipe becomes hot in the sunlight while the rest stays cool in the shade a difference in water level will occur so this situation must be avoided. In sub-zero conditions anti-freeze should be added to the water.

The water level is simple in operation and can be used around corners. The accuracy that can be expected is ±5 mm over a distance of 15 m. Floors and ceilings can be marked out by holding a measured rod against the level of water. A water level is shown in Fig. 7.9.

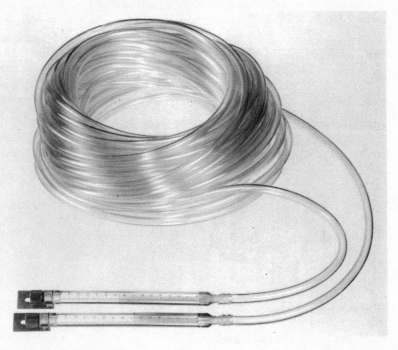

Fig. 7.9 Stabila water level

Mirror automatic level

This level is easy to use without previous knowledge of reducing levels. It contains 2 sets of mirrors, one set giving an erect image while the other set causes the image to appear inverted. It is automatic in that, once set up, a pendulum mirror maintains a horizontal sight line. To prevent the pendulum mirror being damaged in transit it is locked in position when the instrument is not in use.

As the instrument contains mirrors and not lenses there is no magnification. Maximum lengths of sights in the region of 25 m provide an accuracy of ±5 mm.

Method of use

1. Set up the tripod on firm ground with the tripod pin pointing upwards.
2. Push the level on the tripod pin.
3. Adjust the legs until the 2 half circles seen in the top aperture of the Cowley level meet to form one complete circle. When using the Contract level press the button to release the pendulum lens and adjust the legs so that a complete square is seen.
4. Signal the assistant to raise or lower the target on the staff until it appears in coincidence as one line.
5. The assistant clamps the target in position and reads its height on the graduated staff.

A bricklayers' wall stand replaces the tripod when the Cowley level is used on flat surfaces. The Contract level will stand on its own base. It is

Fig. 7.10 Cowley automatic level

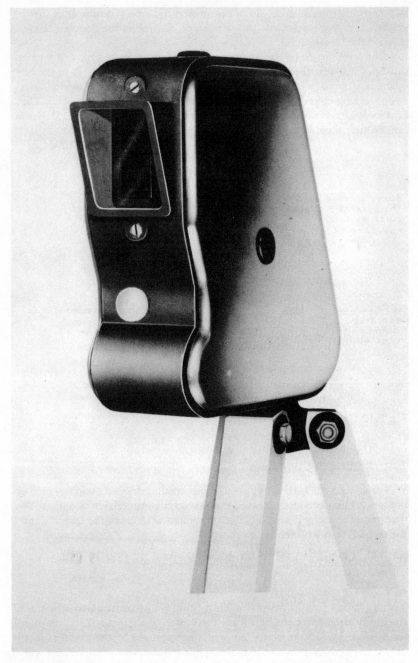

Fig. 7.11 Contract super level

130

important that the level should not be moved to a new position while the pendulum mirror is released. This is easier to control on the Contract level than on the Cowley level.

The level should be checked for accuracy from time to time and for this the 2-peg test described in 'Permanent adjustments to level', p. 76 in Chapter 4, and illustrated in Fig. 4.20 should be used. If the instrument is found not to be working correctly it should be returned to the supplier for repair. The Cowley level is shown in Fig. 7.10 and the Contract level in Fig. 7.11. Operation of both these levels is shown in Fig. 7.12.

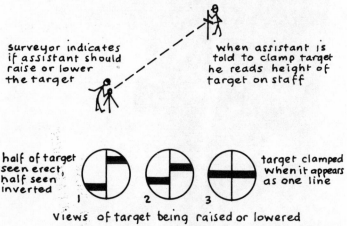

Fig. 7.12 Use of mirror automatic level

7.4. Setting-out buildings

This deals with the setting-out involved for small unframed buildings such as houses and blocks of flats. Setting-out is always considered in two parts. The first part is concerned with the plan or horizontal shape of the building and the second deals with the vertical shape.

Horizontal setting-out

Building line

This is a line in front of which buildings must not be constructed. The Local Authority decides its position from the centre of the road. The drawings showing the building to be set-out would be approved by the Local Authority only if they complied with the building line requirement so this aspect will already have been dealt with by the time setting-out takes place.

Base line

The drawings will show the position of the proposed building dimensioned from existing site features. The surveyor draws a base line on the plan which will be set-out on site and from which the building can be positioned. For a rectangular building the base line would normally be the front line of the building. Where the building is angled to the road the base line position may have to be calculated.

Degree of accuracy

BS 5606:1978 states that the permissible deviation for horizontal brick walls up to 40 m in length is ±40 mm. This required accuracy of 1/1000 can be achieved using a steel tape held horizontally and supported to avoid sag. Adjustment for temperature variation is not necessary.

Pegging-out

All points are marked by wooden pegs with a nail accurately positioned on the top. Measurements are taken with the end loop of the steel tape over the nail. Such measurements are made horizontally with the tape supported at intervals of not more than 10 m to prevent undue sag. With sloping ground

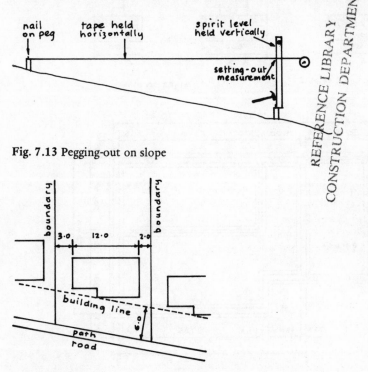

Fig. 7.13 Pegging-out on slope

Fig. 7.14 Block plan

the horizontal measurements must be transferred vertically to the lower peg or board and for this a spirit level is normally used. This is shown in Fig. 7.13.

The horizontal setting-out described below will apply to one particular building. Figure 7.14 is the block plan and shows a detached house with dimensions relating it to the road and boundary. Figure 7.15 is the foundation drawing showing the plan of the foundation walls with the wall sizes.

There are three different forms of wall construction; these are shown in the sections which give the concrete widths and also vertical sizes. Taking information from the plan and sections the complete house foundation can be set-out. Setting-out for the building is described in sequence, each stage being shown in Fig. 7.16.

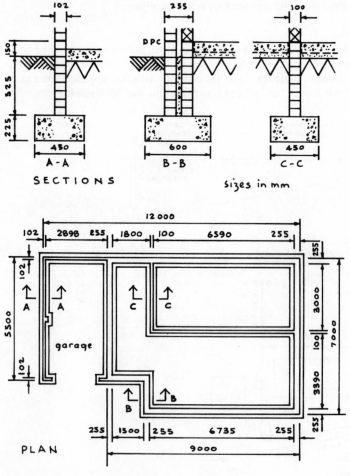

SECTIONS

Sizes in mm

PLAN

Fig. 7.15 Foundation drawing

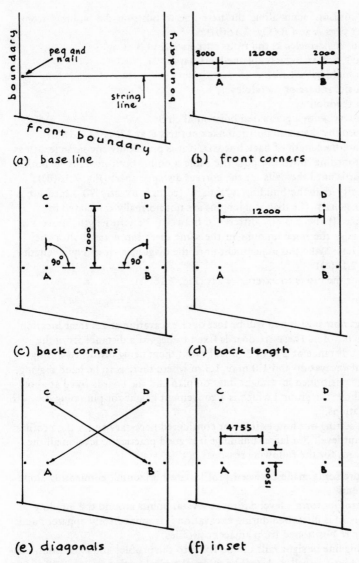

Fig. 7.16 Stages in setting-out a building

Stages
1. Position the base line by measuring from existing site features. In the absence of required site dimensions these can be calculated as in question 7.1 or scaled from the drawing, set-out approximately, then checked and adjusted to conform with given sizes. The line is marked by pegs with nails in the top and string line between them. It is of no particular length but must be longer than the building (Fig. 7.16(a)).

2. Take measurements along the base line to position the building front corner pegs A and B (Fig. 7.16(b)).
3. Set-out right-angles at the front corners A and B using:
    (a) a 3, 4, 5 triangle (explained in Chapter 2);
    (b) a builders' square;
    (c) a site square or theodolevel;
    (d) a theodolite;
    (e) a level with a graduated horizontal circle.
4. Measure the sides and position back corners C and D (Fig. 7.16(c)).
5. Measure the length of back line CD. If it is practically the same length as the front line move the pegs the same amount both inwards or both outwards until the nails are the correct distance apart (Fig. 7.16(d)).
6. To verify that the building is 'square' (corners exactly 90°) measure the diagonals. The diagonal lengths are not normally calculated in advance. If there is any difference between the two lengths move the back pegs the same amount in the same direction parallel with the front line. Make this adjustment until the diagonals are of equal length (Fig. 7.16(e)).
7. Peg-out the insets to external walls (Fig. 7.16(f)).

Profiles

As the pegs now positioned will be lost once excavation starts their location is fixed by profiles. These are boards fixed to pegs at a distance from the building. A distance of 2 m or 3 m will avoid them being disturbed during mechanical excavation and 1.0 m or 1.5 m where there is to be hand digging. Profiles are positioned in straight lines on plan and the boards fixed at about 300 mm above the ground which is a convenient height for plumbing down to the foundations.

When setting-out houses it is not considered necessary to fix the profiles at a constant level. For larger buildings it is good practice to level-in all the profile boards for the following reasons:

(a) Measurements made between profiles are horizontal, eliminating slope problems.
(b) The profiles form a level datum at several points around the building.
(c) If a board is disturbed during excavation it is immediately apparent and can be re-positioned from adjacent profiles.
(d) A string line or sight rails can be fixed to the profiles and excavation controlled by a traveller, described under 'Vertical setting-out', p. 138.

Corner profiles. When the profiles at the corners of the rectangle are positioned the line of the external wall face is marked on the boards. This is done by stretching a string line from the nail in one corner peg to pass over the top of the nail in the other corner peg and extend to the profile where it is marked (see Fig. 7.17). When all the corner profiles have been marked the corner pegs can be removed.

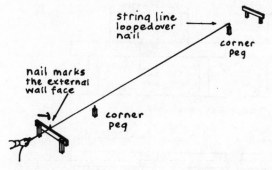

Fig. 7.17 Transferring building corner points to corner profiles

Intermediate profiles. Between the corner profiles the intermediate profiles
are fixed in a straight line. They mark the position of insets in the external
wall and also any internal walls which have foundations. Internal walls built
off the concrete ground floor slabs are not set-out with profiles.

The intermediate profiles are marked out and positioned by using running measurements from the corner profiles. Running measurements are not
shown on drawings and must therefore be calculated from the separate
measurements given.

Once the profile marks have been shown they should be checked using
the separate measurements from the drawing.

Profile marking

The marks are usually made by driving wire nails in the top of the profile
boards, alternatively saw cuts can be used. The purpose of each mark must be
written on the board with paint or crayon. Another method which shows up
distinctly is to paint the width of the wall on the face of the board. When the
walls are set-out from the profile boards, one string line is used for the face of
a solid wall and two lines for the faces of a cavity wall.

Where foundations are to be dug by hand the total width of the concrete
is shown on the profiles. For mechanical excavation usually only the centre-
line is marked, a bucket of the appropriate size being fitted to the excavator
to suit the trench width (see Fig. 7.18).

Figure 7.19 shows the foundation plan with profiles in position. Note
that where different walls are closely spaced they are marked on the same
profile.

Marking out trenches

The topsoil may be stripped off before or after the profiles are fixed.

String lines are stretched between foundation concrete marks on opposite
profile boards. Dry cement, dry lime or sand is sprinkled on the string line
and falls to the ground forming a straight line. Only that part of each string
line necessary to show the shape of the foundation trenches is marked out.
The string lines are removed and excavation can take place. See Fig. 7.20.

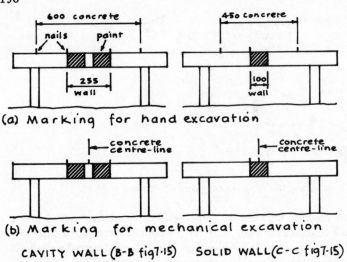

(a) Marking for hand excavation

(b) Marking for mechanical excavation

CAVITY WALL (B-B fig7.15)   SOLID WALL (C-C fig7.15)

Fig. 7.18 Profile marking

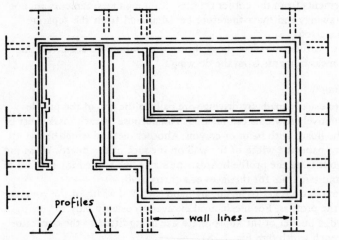

Fig. 7.19 Position of profiles

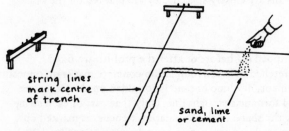

Fig. 7.20 Marking trenches

Offset pegs in place of profiles

This sytem uses a single peg instead of each profile board, the peg being off-set from the building by 2 or 3 m. It is explained as a sequence of operations.

1. Peg-out the shape of the building as already described under 'Pegging-out' on p. 131, including all stages shown in Fig. 7.16.
2. Position the offset pegs by a string line extended from the building pegs using a nail to show the accurate position.
3. Remove the building pegs.
4. Stretch a string line between opposite offset pegs.
5. Standing above this line use a tape or marked batten to position on the ground a second string line to one side and parallel with the first string line. This marks the centre-line of the excavation. Alternatively two string lines on the ground may be used to mark the edges of the excavation.
6. Wrap the string line around a brick at each end to keep it in position.
7. Mark the line with dry cement, lime or sand. Refer to Fig. 7.21.

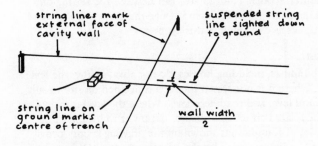

Fig. 7.21 Offset pegs in place of profiles

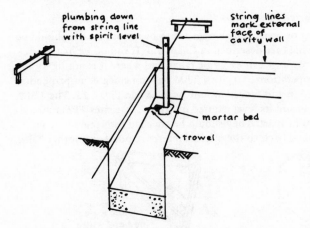

Fig. 7.22 Marking foundation walls

Comparison between profile boards and offset pegs

Once profile boards are fixed and marked it is easy, at any time, to mark out the foundation trenches using only string lines. However, for a simple foundation plan the offset peg method takes less time to set-out but skilled setting-out is necessary during the excavation marking-out stage.

Marking out walls

After the concrete has been poured and set, the walls are set-out. String lines are stretched between wall face marks on opposite profile boards or offset pegs. A spirit level is used to plumb down from the string line to the mortar bed where a line is marked with a trowel. See Fig. 7.22. The string line is then removed.

The brick or block corners are raised first and a straight face line is maintained by using a string line for each course. Verticality is maintained by using a spirit level.

Setting-out by polar co-ordinates

As explained in Chapter 1 polar co-ordinates can be used for setting-out corners as shown in question 5.5 in Chapter 5. Once the corners are marked out setting-out continues as explained in this chapter.

Vertical setting-out

Drawings for small buildings, including houses, do not always show the level of the ground floor. Where it is not stated the level is decided on site to suit the surrounding ground level and road kerb level. Where the level is shown on the drawing it will be related either to ordnance datum or to local datum.

The depth or level of foundations shown on the drawing is only provisional. The final depth will be decided by the Building Control Officer during his inspection of the trench.

Temporary or transferred bench-mark

Where a large building or several small ones are to be set-out a TBM should be established on site. TBMs are described in Chapter 4. It should be in a prominent position where readings can easily be taken and where it is not likely to be disturbed by site operations. To give a TBM peg some degree of protection it should be concreted in and fenced around as shown in Fig. 7.23. The TBM should be painted blue and its level painted on it. On large sites TBMs should be not more than 100 m apart. They should be checked for accuracy from time to time by levelling back to the original reference point or another TBM.

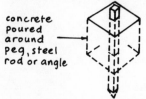
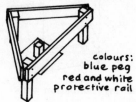
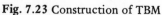

concrete poured around peg, steel rod or angle

colours: blue peg red and white protective rail

Fig. 7.23 Construction of TBM

Building datum

A reference surface must be established close to the building being set-out. This datum should be at a particular level on the building. DPC or floor slab level (often the same) is more convenient for working from than the finished floor level which is sometimes used.

Where profile boards are fixed at the same level around the building that level should be datum level. On all other occasions pegs are driven in at a distance of not more than 0.6 m from the edge of the trench excavation. A peg at each corner is adequate for small buildings with the floor slab at one height. Where there is a change of level in the floor slab four corner pegs should be positioned for each floor level.

Foundation depth

As the work proceeds foundation trenches are checked at intervals for depth of excavation. The bottom of the trench is horizontal but will be stepped on sloping sites. The depth of excavation can be controlled by one of the following five methods.

Hand digging

1. Where profiles are at the same level a string line can be stretched between them and a travelling rod used to measure down to the trench bottom. This is shown in Fig. 7.24(a).

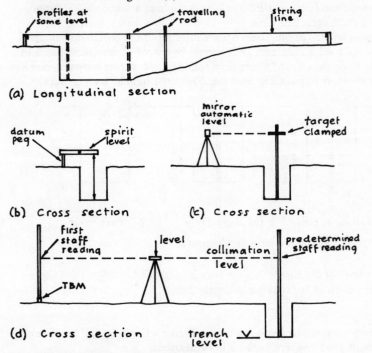

(a) Longitudinal section

(b) Cross section

(c) Cross section

(d) Cross section

Fig. 7.24 Foundation trench depth control

2. A spirit level can be held on a datum peg so that it extends over the trench and a vertical measurement can then be taken. The bottom of the trench is kept horizontal by using the spirit level on a straight edge board. A check is made when the next datum peg is reached (see Fig. 7.24(b)). For mechanical excavation this method is too laborious and the datum pegs are likely to be disturbed.

Mechanical excavation

3. By use of a mirror automatic level. Clamp the target on the staff at the appropriate reading to give the required depth of trench (see Fig. 7.24(c)).

4. By use of level and staff. Calculate the appropriate staff reading to give the required depth of trench by subtracting the trench level from the collimation level. *Example:* TBM 24.62 + staff reading 1.54 = collimation level of 26.16 − trench level of 23.50 gives required staff reading of 2.66. See Fig. 7.24(d).

5. By use of site rails and traveller as described later under 'Use of sight rails for foundation trench excavation' on p. 142.

Top of foundation concrete

The thickness of foundation concrete shown on the drawings should be treated as a minimum and can be increased to suit brick or block courses being built from it up to DPC level. Brick courses are normally 75 mm including the mortar bed joint.

Pegs or lengths of mild steel bar are driven into the trench bottom until the required level is obtained. They should be spaced at approximately 3 m intervals along the trench. Where deep foundation concrete is used it may be more convenient to push pegs into the side of the trench. See Fig. 7.25.

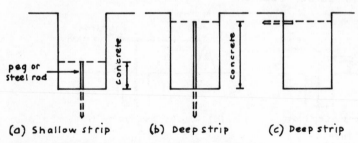

Fig. 7.25 Foundation concrete depth control

The level of the tops of pegs or bars is controlled by any of the five methods described for excavation depth control.

Safety

The surveyor should be familiar with the Construction (General Provisions) Regulations 1961 concerning work in excavations.

7.5. Setting-out for bulk excavation

This deals with controlling the excavation required to level-off part of a site before the building setting-out is done.

Horizontal setting-out

The surface is stripped off and corner pegs positioned. When measuring up or down slopes allowances must be made to obtain the true horizontal length. Right-angles are set-out using any of the methods previously described for setting-out buildings. String line is stretched between corner pegs and dry cement, lime or sand is used to mark the line on the ground.

Vertical setting-out

The finished surface level after excavation has been carried out is called the formation level. The principle adopted is to set-up a sight line above the ground which is parallel to the formation line. This is done by fixing sight rails at opposite ends of the excavation with one or both of them at a convenient height above the ground for sighting over. Between the sight rails a wooden gauge called a traveller is held vertically. Its length is such that when the top is in line with the sight rails the bottom is at formation level.

Sight rails

These are nailed horizontally to upright stakes at a fixed distance from the edge of the excavation using a spirit level. The ones for sighting from should be between 1.0 m and 1.5 m above the ground. They are fixed in pairs at opposite ends of the excavation and there must be a minimum of two pairs to enable the whole site to be covered and to act as a check on accuracy (see Fig. 7.26). If painted they should have black and yellow vertical stripes.

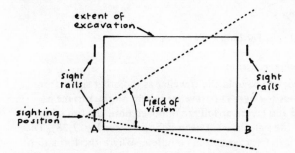

Fig. 7.26 Plan of sight rails for bulk excavation

Traveller

A traveller is made from wood in the shape of the letter 'T'; it is normally made a multiple of 0.25 m in length to suit the depth of excavation. A stand can be added to the base to make it self-supporting. If painted it should have black and yellow stripes like the sight rails.

142

Setting-up sight rails

The example calculations given here refer to Fig. 7.27. Formation level is 32.63. Peg A is positioned 2.0 m away from the excavation and driven in practically to ground level. Site rail stakes are set-up on either side of the peg. By staff readings taken from the site TBM, the peg A is found to be 33.24 OD.

For sight rails to be between 1.0 m and 1.5 m above the ground the top of the sight rail at A will be between 34.24 and 34.74 OD. The difference between the minimum height of 34.24 and the formation level of 32.63 is 1.61 m. For the length of the traveller to be a multiple of 0.25 m it must be 1.75 m long. Once the length of the traveller has been decided the level of the sight rail at A is found by adding the travelling length to the formation level. 32.63 + 1.75 = 34.38 OD. As the formation level at B is the same as at A the sight rail level at B will also be 34.38 OD.

The height of the top of the sight rail above peg A is 34.38 − 33.24 = 1.14 m. This is measured up from A by using a spirit level to mark the peg level on one of the stakes and measuring up the stake with a tape or the staff. The sight rail is nailed to the stakes using a spirit level to ensure that it is horizontal.

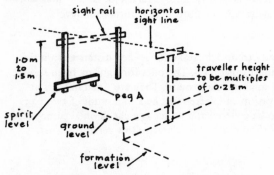

Fig. 7.27 Height of sight rails for bulk excavation

Use of sight rails for bulk excavation

Two people are required, one to hold the traveller as the other sights from one sight rail to the opposite ones. The traveller is held on the surface of the soil being excavated and layers of soil are removed until the top of the traveller coincides with the sight line between sight rails (see Fig. 7.28). The process is repeated until the excavation is complete. Where the formation level slopes the sight rails are fixed at the same slope.

Use of sight rails for foundation trench excavation

Sight rails and traveller can also be used for foundation trenches. It is not normally considered worthwhile to set them up for small buildings but they are used for larger setting-out work, particularly if they are already in position for bulk excavation purposes.

Where the bottom of the excavation is stepped different length travellers would be needed.

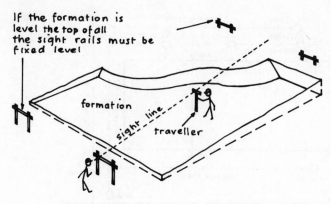

Fig. 7.28 Use of sight rails for bulk excavation

7.6. Setting-out drains

The terminology used in drainage is illustrated in Fig. 7.29 and drain slope calculations in Fig. 7.30.

Drainage drawings will show the following information:
1. diameter of drains;
2. whether for soil, surface water or a combined system;
3. direction of flow and gradient;
4. invert levels at manholes.

Small buildings are likely to have this information shown on a site plan. An example is given in Fig. 7.31.

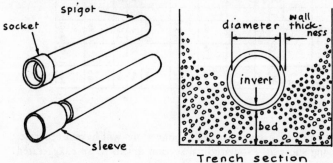

Fig. 7.29 Drainage terms

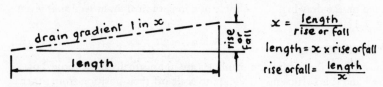

Fig. 7.30 Drain slope calculations

$$x = \frac{length}{rise\ or\ fall}$$

$$length = x \times rise\ or\ fall$$

$$rise\ or\ fall = \frac{length}{x}$$

144

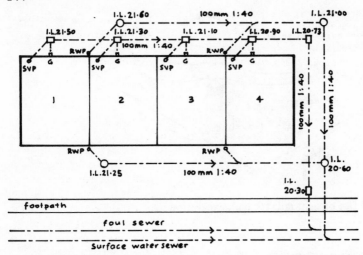

I.L. Indicates the invert level above Ordnance Datum Newlyn

Fig. 7.31 Drainage plan for block of 4 houses

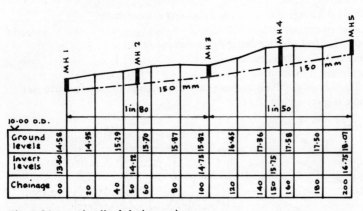

Fig. 7.32 Longitudinal drain section

Larger schemes with several branch connections will have a separate drainage drawing showing a longitudinal section drawn to an exaggerated scale. An exaggerated scale has the vertical scale larger than the horizontal scale and its purpose is to show the gradients clearly. A plan may also be included on the drawing. An example of a longitudinal drain section is given in Fig. 7.32.

Horizontal setting-out

The position of drain runs and manholes are not normally dimensioned on the plan and it is sufficiently accurate to scale off their position from other site features. The centre-line should be marked on the ground by pegs painted white. The maximum distance between pegs is 30 m with a peg at the centre of each manhole. As the centre-line pegs will be lost during excavation, wherever they occur offset pegs are fixed at a distance of 2 m or 3 m to one side of the centre-line. The offset distance must be the same between any two manholes. Offset pegs are painted yellow.

Vertical setting-out

The gradient of drains must be strictly controlled. The Maguire rule of 1 in 40 for a 100 mm diameter, 1 in 60 for a 150 mm diameter and 1 in 90 for a 225 mm diameter pipe is frequently used but less inclined slopes are still self-cleansing. Considerable accuracy in setting-out and in pipe laying is necessary in order to maintain such minimal slopes and so avoid blockages when the drains are used.

The level of the point of discharge should first be checked for accuracy by taking levels. If the drawing shows an incorrect existing invert level the gradients and inverts shown on the drawing will have to be recalculated.

Sight lines

The method used to control the depth of excavation and the invert level of the drain pipes being laid is to set-up a sight line above the ground, parallel to the gradient of the drain. A traveller is used between fixed sight rails.

When hand-digging is employed the sight rails can straddle the trench and a 'T'-shaped traveller can be used. When the trench is dug by machine the excavator works on top of the trench centre-line thus obscuring sight rails set-up on the trench line. In this situation the sight rails have one upright stake and are positioned to one side of the trench and a traveller shaped like an inverted letter 'L' is used to reach the drain line.

The sight rails are positioned next to the offset pegs and so will have the same spacing. In a straight run there should be a minimum of three sight rails so that if one is disturbed it is immediately evident and can be set-up again by sighting from the other two.

When setting-up sight rails on a straight run each one must be set-up independently from the others using levels from the drawing or calculated levels. Their final lining-up by sight provides a check on accuracy.

To be at a convenient height for sighting over, the rails should be between 1 m and 1.5 m above the ground. The traveller is made a multiple of 0.25 m from the top to the steel invert bracket at the bottom.

Setting-up sight rails

The calculations refer to the drain between MH1 and MH2 in Fig. 7.32.

Sight rails should be set-up at a maximum distance of 30 m. As MH1 and MH2 are 50 m apart sight rails will be set-up at chainage (ch) 00, chainage 25 and chainage 50. The stages in setting-up and using the sight rails in this example are shown in Fig. 7.33. By interpolating between the drain invert level at ch 00 and ch 50 the invert level at ch 25 is found to be 13.81. Staff readings from the site TBM to the top of the offset pegs give levels of 14.64 at ch 00 and 15.12 at ch 25. See Fig. 7.33(1).

146

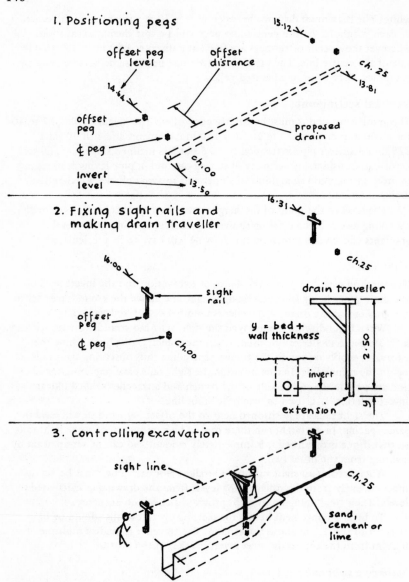

Fig. 7.33 Drainage sight rails

For the sight rails to be 1.00 m to 1.50 m above the ground, and treating the top of the offset pegs as ground level, the sight rail at ch 00 must be between 15.64 and 16.14 and at ch 25 between 16.12 and 16.62. The difference between the minimum sight rail level and the drain invert level at ch 00 is 15.64 − 13.50 = 2.14 and at ch 25 is 16.12 − 13.81 = 2.31.

The traveller must be in multiples of 0.25 m so a length of 2.5 suits both chainage points giving a sight rail height at ch 00 of 13.50 + 2.50 = 16.00, and at ch 25 of 13.81 + 2.50 = 16.31. See Fig. 7.33(2).
The results are shown in tabular form.

Chainage	Invert level	Traveller length	Sight rail level	Top of offset peg	Sight rail above peg
00	13.50	2.50	16.00	14.64	1.36
25	13.81	2.50	16.31	15.12	1.19

The height of the sight rail above the offset peg is measured with a tape or staff. The sight rail is nailed to the stake and a spirit level used to ensure that it is horizontal.

Use of sight rails for drain excavation
An allowance for the depth of the bed and thickness of pipe is made by adding a short extension piece to the bottom of the traveller shown as 'y' in Fig. 7.33(2).
Depth of excavation is checked at intervals along the trench as the work proceeds, the top of the traveller being made to coincide with the sight line between one sight rail and the next. See Fig. 7.33(3).

Use of sight rails for drain laying
The extension piece is removed from the traveller. All invert levels are found by means of a steel bracket screwed to the foot of the traveller. Drain laying starts at the lower level and continues with the sockets facing up the slope. Three different ways to align the pipes to the required gradient are described.

1. Aligning pipes direct from invert bracket. The granular bed is poured into the trench and pipes are laid on top. Each pipe is aligned by the invert bracket touching the pipe invert while the traveller crosspiece is on the sight line (see Fig. 7.34(a). This method is satisfactory when using long lengths of pipe. If used for short pipes it is doubtful if all of them will be to exactly the same gradient.

2. Aligning pipes by invert level pegs. Pegs are driven into the bottom of the trench at 3 m intervals. The level of each one is fixed by the foot of the traveller while the traveller crosspiece is on the sight line. The bed is poured into the trench and the pipes laid on top. A 3 m straight edge fits inside the pipe and extends to the next invert peg. Pipes are laid in turn keeping their inverts touching the straight edge. When the pipes reach a peg it is removed. This method is shown in Fig. 7.34(b).

3. Aligning pipes by string line. With this method pegs are driven into the trench bottom at intervals and, measuring up from the traveller bracket at each peg, the string is fixed at the top of the pipe sockets or sleeves.

148

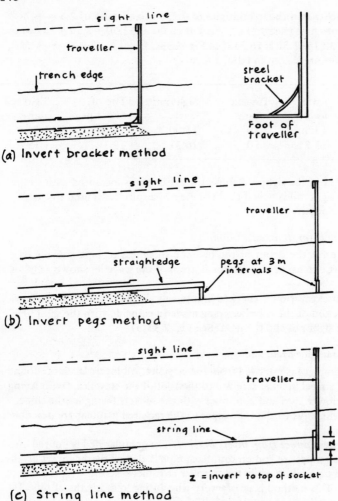

(a) Invert bracket method

(b). Invert pegs method

(c) String line method

Fig. 7.34 Methods of aligning drain pipes

The string line is tied tightly between a fixed pipe and the next peg and the pipes are laid so that their sockets or sleeves just touch the string. It is easier to see if the pipes are correctly aligned if the string is kept a small distance above the pipe joints.

The disadvantage of this method is the inevitable sag in the string. The pegs should not be more than 5 m apart. See Fig. 7.34(c).

Double sight rails

Double sight rails are used either because the bottom of the trench is stepped in cross-section or that the drain does not follow the natural slope of the ground.

1. *For use with dual drains.* Where trenches are to accommodate two drains or other services at different levels the solution is either to use double sight rails as in Fig. 7.35(a) or, if the difference in trench level is constant, to use a double traveller. A double traveller has two crosspieces to suit the different depths of excavation and is shown in Fig. 7.35(b).

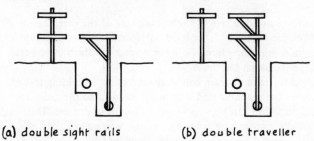

(a) double sight rails   (b) double traveller

Fig. 7.35 Dual drain trenches

2. *For use with backdrop manholes.* The purpose of a backdrop manhole is to accommodate a change in drain invert levels. To control differing depths the same traveller is used but two sight rails are fixed to the same stake. The difference in height of the sight rails corresponds to the difference in level of the incoming and outgoing drain inverts. See Fig. 7.36(a).

3. *For use with uneven ground slopes.* A fixed length traveller can be used only where the drain gradients roughly follow the ground slope. Any variation will cause the drain to be at a varying depth below ground level. To keep the sight rails between 1.0 m and 1.5 m above uneven ground different length travellers will have to be used. This is illustrated in Fig. 7.36(b).

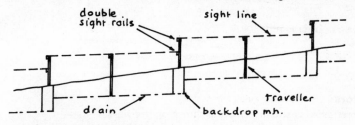

(a) backdrop manholes

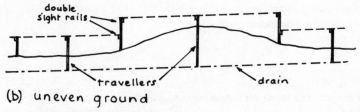

(b) uneven ground

Fig. 7.36 Uses of double sight rails

150

Wherever different length travellers are used on the same drain run the double sight rails must be marked accordingly. The length of the appropriate traveller should be shown on the sight rail facing the direction in which the traveller is to be used.

Boning-in drains

Boning rods are made of wood in sets of three, each one of the three being the same height. They are normally 'T'-shaped but special ones of different shapes and colours incorporating plumb-bobs for greater accuracy have been designed by the Road Research Laboratory.

They can be used for branch drain connections from the building to the manhole. If the pipe at the manhole end is in position and also the gully at the building end a boning rod can be positioned on top of each pipe. The drain pipes between the two can be fixed in line by 'boning-in', that is, standing the third boning rod on each pipe being fixed and sighting through so that the top of all three rods are in line. This method is illustrated in Fig. 7.37.

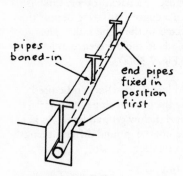

Fig. 7.37 Use of boning rods

7.7. Errors

Gross errors

(a) Incorrect existing site information shown on drawings.
(b) Incorrect setting-out information shown on drawings.
(c) Incorrect measurements made on site.
(d) Incorrect use of instruments.
(e) Use of disturbed reference points.

Remedy

(a) 'Prove' the setting-out drawing for existing measurements and levels.
(b) Check drawing calculations.
(c) Once lengths and levels are set-out on site, check them by a different method.
(d) Verify the accuracy of reference points from time to time.

Systematic errors

These are caused by errors in instruments.

Remedy

(a) Check measuring instruments against a standard length as described under 'Standardizing', p. 20 in Chapter 2.
(b) Check squares in two directions from a straight line.
(c) Check levels as described under 'Permanent adjustments', p. 75 in Chapter 4.

7.8. Questions

Formulae shown in Appendix 1 (p. 154).
Qu 7.1 Figure 7.38 shows certain setting-out dimensions for a house. Using trigonometry calculate the lengths 'a' and 'b'.

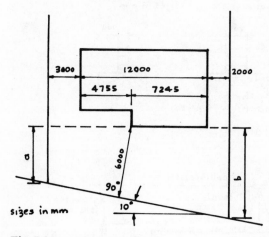

Fig. 7.38

Qu 7.2 The foundation walls shown in Fig. 7.39 are for a pair of houses. Copy the plan and on it show where profile boards should be positioned. Using broken lines indicate the wall faces that need to be shown on the profile boards for setting-out purposes.

Qu 7.3 Figure 7.40 is a section through a foundation trench. With the level as shown and the staff positioned on top of the points to be checked, calculate the staff readings that will show that the following points are at the required level:

(a) bottom of trench;
(b) top of concrete foundation peg;
(c) top of datum peg.

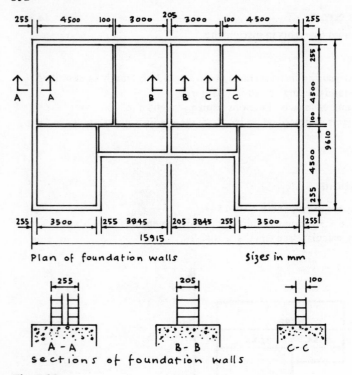

Plan of foundation walls   Sizes in mm

sections of foundation walls

Fig. 7.39

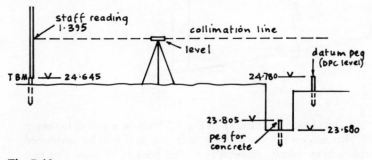

Fig. 7.40

Qu 7.4 This question on setting-out for bulk excavation relates to Fig. 7.27. All levels relate to OD Newlyn. By levelling from the site TBM the level of peg A was found to be 44.36. The required horizontal formation level is 43.45. Calculate:

(a) suitable traveller of 0.25 m multiples;
(b) level of sight rail;
(c) difference between level of peg A and the sight rail.

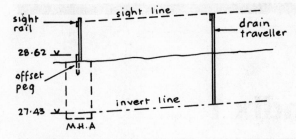

Fig. 7.41

Qu 7.5 Figure 7.41 shows vertical setting-out arrangements for a drain. The horizontal length between MHA and the other end of the drain at MHB is 84.50 m. The invert at MHB is 29.12 m. It is required to erect sight rails at 30 m and 60 m from MHA on the drain run AB.

(a) Maintaining a traveller length multiple of 0.25 m calculate a suitable traveller length.
(b) Find the gradient of the drain.
(c) Prepare a table showing the invert and sight rail levels at MHA, the 30 m and 60 m positions.

Answers shown in Appendix 2 (p. 156).

Appendix 1

FORMULAE

Transposition of formula

example: $a = \dfrac{b}{c}$ To find b $a \cancel{=} \dfrac{b}{c}$ $ac = b$

To find c $a \cancel{=} \dfrac{b}{c}$ $c = \dfrac{b}{a}$

Length of triangle sides

Pythagoras theorem

$$c^2 = a^2 + b^2 \qquad c = \sqrt{a^2 + b^2}$$
$$b^2 = c^2 - a^2 \qquad b = \sqrt{c^2 - a^2}$$
$$a^2 = c^2 - b^2 \qquad a = \sqrt{c^2 - b^2}$$

Triangle with sides of 3, 4, 5 units contains a 90° angle

Trigonometry

$$\sin \theta = \dfrac{opp}{hyp}$$
$$\cos \theta = \dfrac{adj}{hyp}$$
$$\tan \theta = \dfrac{opp}{adj}$$

Sine rule $\dfrac{a}{\sin A} = \dfrac{b}{\sin B} = \dfrac{c}{\sin C}$

Cosine rule

$$a^2 = b^2 + c^2 - 2bc \cos A \qquad \cos A = \dfrac{b^2 + c^2 - a^2}{2bc}$$
$$b^2 = c^2 + a^2 - 2ca \cos B \qquad \cos B = \dfrac{c^2 + a^2 - b^2}{2ca}$$
$$c^2 = a^2 + b^2 - 2ab \cos C \qquad \cos C = \dfrac{a^2 + b^2 - c^2}{2ab}$$

Angles

Sexagesimal division used in Great Britain
Whole circle = 360 degrees(°)   1° = 60 minutes(')   1' = 60 seconds(")

Centesimal division used in parts of Europe
Whole circle = 400 grades(ᵍ) Divisions of grades in decimals

Changing minutes and seconds to decimal of a degree
73° 08' 42",   42÷60 = 0.7,   73° 08.7',   08.7÷60 = 0.145,   73.145°

Changing decimal of degree to minutes and seconds
73.145°,   0.145 × 60 = 8.7,   73° 08.7',   0.7 × 60 = 42,   73° 08' 42"

Slopes

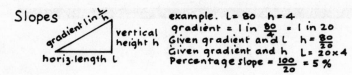

gradient 1 in $\frac{L}{h}$

vertical height h

horiz. length L

example. L = 80 h = 4

gradient = 1 in $\frac{80}{4}$ = 1 in 20

Given gradient and L   h = $\frac{80}{20}$

Given gradient and h   L = 20 × 4

Percentage slope = $\frac{100}{20}$ = 5%

Areas   1 hectare (ha) = 1000 m^2

Areas of triangles

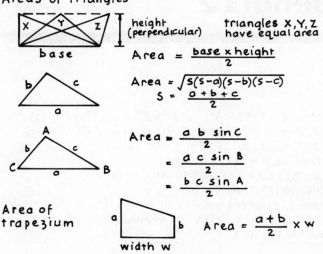

X   Y   Z

base

height (perpendicular)

triangles X, Y, Z have equal area

$$Area = \frac{base \times height}{2}$$

$$Area = \sqrt{S(S-a)(S-b)(S-c)}$$

$$S = \frac{a+b+c}{2}$$

$$Area = \frac{a\,b\,\sin C}{2}$$

$$= \frac{a\,c\,\sin B}{2}$$

$$= \frac{b\,c\,\sin A}{2}$$

Area of trapezium

a   b   width w

$$Area = \frac{a+b}{2} \times w$$

Areas from offsets

1. Trapezoidal rule. Applies to any number of equal width strips

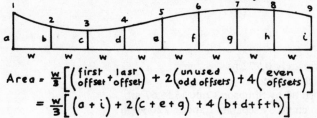

a   b   c   d   e   f   g   h   i

w   w   w   w   w   w   w   w

$$Area = w\left[\left(\frac{first\ offset + last\ offset}{2}\right) + \frac{remainder}{of\ offsets}\right]$$

$$= w\left[\left(\frac{a+i}{2}\right) + b+c+d+e+f+g+h\right]$$

2. Simpson's rule. Applies to an even number of equal width strips enclosing a curve

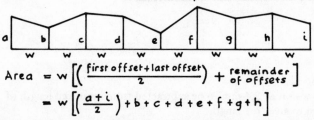

a   b   c   d   e   f   g   h   i

w   w   w   w   w   w   w   w

$$Area = \frac{w}{3}\left[\left(\substack{first \\ offset} + \substack{last \\ offset}\right) + 2\left(\substack{unused \\ odd\ offsets}\right) + 4\left(\substack{even \\ offsets}\right)\right]$$

$$= \frac{w}{3}\left[(a+i) + 2(c+e+g) + 4(b+d+f+h)\right]$$

Appendix 2

Answers

Chapter 1

Qu 1.1 AC = 87.70, BC = 175.54, AD = 157.75, BD = 104.75, CD = 149.78

Qu 1.2 C $20°$ 18.79 m, D $40°$ 25.32 m, E $60°$ 28.79 m, F $80°$ 28.79 m, G $100°$ 25.32 m, H $120°$ 18.79 m, J $140°$ 10.00 m

Qu 1.3 Offsets D = 6.42, E = 8.64, F = 12.06
AH = 6.05, AJ = 13.40, AK = 15.50, AL = 20.15

Qu 1.4 (a) 1/1200, (b) 137.5 mm

Qu 1.5 152.4 mm reps 1 609 344 mm = 1 : 10 560
1 km = 1 000 000 mm ÷ 10 560 = 94.697 mm
Scale shown in Fig. A2.1

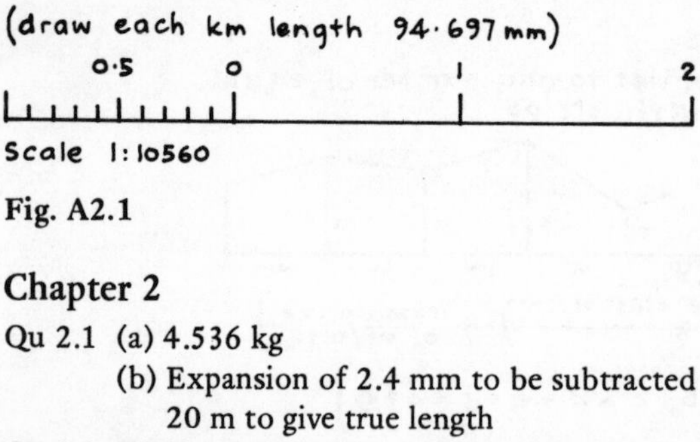

(draw each km length 94·697 mm)

Scale 1 : 10560

Fig. A2.1

Chapter 2

Qu 2.1 (a) 4.536 kg
(b) Expansion of 2.4 mm to be subtracted from measured length of 20 m to give true length

Qu 2.2 0.541 error, 216.4 − 0.541 = 215.859 true length

Qu 2.3 (a) 1 in 4.51 gradient, (b) 22.17 per cent slope, (c) 48.03 m

Qu 2.4 Fig. A2.2

Qu 2.5 Fig. A2.3

Chapter 3

Qu 3.1 Approximately 0.772 ha

Qu 3.2 Left of chain line:
ch 0 to ch 80 by Simpson's rule 1534.67 m^2
ch 80 to ch 96 by separate trapezia 220.90 m^2

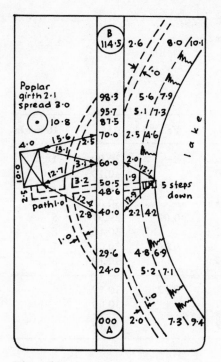

Fig. A2.2

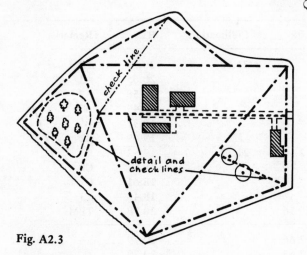

Fig. A2.3

Right of chain line:
ch 0 to ch 90 by trapezoidal rule 1757.00 m²
ch 90 to ch 96 by separate trapezium 102.60 m²
Total area = 0.361 52 ha

Qu 3.3 1.1938 ha
Qu 3.4 0.5081 ha

Chapter 4

Qu 4.1

Table A2.1

BS	IS	FS	Rise	Fall	RL	Remarks
1.16			—	—	30.00	1
	1.06		0.10		30.10	2
	1.25			0.19	29.91	3
	1.12		0.13		30.04	4
	1.19			0.07	29.97	5
	1.04		0.15		30.12	6
	1.10			0.06	30.06	7
		0.97	0.13		30.19	8
1.16		0.97	0.51	0.32	30.19	
0.97			0.32		30.00	
0.19			0.19		0.19	

Qu 4.2

Table A2.2

BS	IS	FS	Collimation	RL	Remarks
0.83			21.62	20.79	TBM
	0.97			20.65	A1
	1.04			20.58	A2
	1.17			20.45	A3
	2.25			19.37	B3
	2.13			19.49	B2
1.82		2.10	21.34	19.52	B1
	2.69			18.65	C1
	2.73			18.61	C2
	2.86			18.48	C3
		0.56		20.78	TBM
2.65		2.66		20.79	
		2.65		20.78	
		0.01		0.01	

Qu 4.3

Table A2.3

BS	IS	FS	Rise	Fall	RL	Remarks
1.343			–	–	36.470	OBM
0.487		1.076	0.267		36.737	
1.007		2.103		1.616	35.121	
1.867		1.942		0.935	34.186	
0.952		0.875	0.992		35.178	TBM
1.763		1.486		0.534	34.644	
2.085		1.329	0.434		35.078	
1.848		1.741	0.344		35.422	
		0.788	1.060		36.482	OBM
11.352		11.340	3.097	3.085	36.482	
11.340			3.085		36.470	
0.012			0.012		0.012	

Permissible closing error = $20\sqrt{k}$ mm, $20\sqrt{0.8}$ = 17.89 mm
To correct the last RL 12 mm must be subtracted. The TBM occurs halfway
round the circuit so correction is – 12/2 mm. 35.178 – 0.006 = 35.172

Qu 4.4

Table A2.4

BS		FS	Collimation	RL	Remarks
1.225			23.475	22.250	A
	1.482			21.993	B
	-2.103			25.578	C
	-2.400			25.875	D
	-1.210			24.685	E
	-1.454			24.929	F
		1.479		21.996	G
1.225		1.479		22.250	
		1.225		21.996	
		0.254		0.254	

Chapter 5

Qu 5.1 AC = 101° 45′ 50″, AD = 218° 20′ 15″, AE = 269° 38′ 27″,
AF = 302° 36′ 14″

Qu 5.2 334.44 m

Qu 5.3 44.31 + 2.0 = 46.31 m

160

Qu 5.4 CAB 70° 19′ 20″, ABC 31° 48′ 00″
Qu 5.5 BC length 21.230 m bearing 16° 25′ 1″
BD length 31.044 m bearing 31° 52′ 18″
BE length 27.898 m bearing 71° 05′ 11″
BF length 16.287 m bearing 79° 13′ 54″

Chapter 6

Qu 6.1 (a) 10.00 m, (b) 3.40 m
Qu 6.2 Fig. A2.4
Qu 6.3 Fig. A2.5

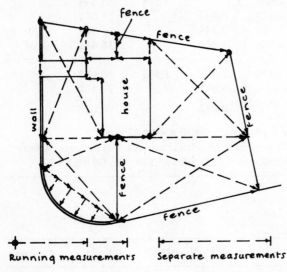

Fig. A2.4

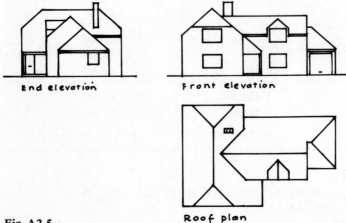

Fig. A2.5

Chapter 7

Qu 7.1 a = 4725 mm, b = 7723 mm

Qu 7.2 Fig. A2.6

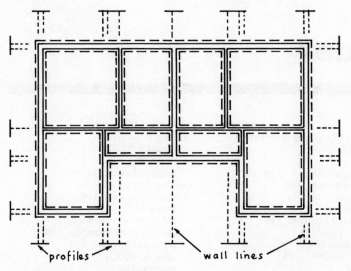

Fig. A2.6

profiles   wall lines

Qu 7.3 (a) 2.460, (b) 2.235, (c) 1.260

Qu 7.4 (a) 2.0 m, (b) 45.45, (c) 1.09

Qu 7.5 (a) 2.25 m, (b)1 in 50

(c)	Position	Invert level	Sight rail level
	MHA	27.43	29.68
	30 m	28.03	30.28
	60 m	28.63	30.88

Index

164